D0949439

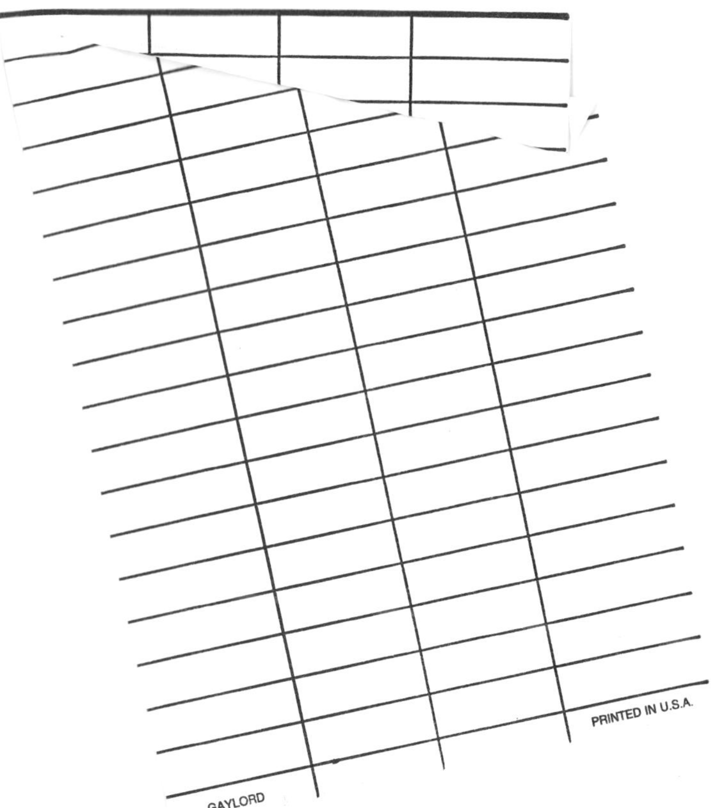
DATE DUE
PRINTED IN U.S.A.
GAYLORD

More praise for *Technology's Promise*

"The pace of change mandates that effective leaders understand the implications and promise that the coming technological tsunami holds. Professor Halal has crafted a prescient masterpiece that takes the reader on a journey to the horizon and back. Filled with poignant and practical insights grounded in data and a remarkable network of experts, *Technology's Promise* moves beyond forecasting and delivers. For Dr Halal fans this book is a wonderful extension of *The Infinite Resource*. Put this book on your wish list and, whatever you do, read it before 2010." – **Lieutenant Colonel Nate Allen**, *PhD, U.S. Army, Co-Founder of CompanyCommand.com, Washington, DC*

"A daring and buoyant excursion into the next few decades, looking at technological change as an integral part of human evolution and human evolution as an integral part of technological change. Should be useful even – maybe especially – for people who will disagree with its forecasts and its essentially hopeful worldview." – **Walter Truett Anderson**, *President Emeritus, World Academy of Art and Science, San Francisco*

"Clearly, business success in the 21st century requires differentiating between the bleeding edge and the leading edge of technological advancements as they morph from innovation to application. In his landmark book, *Technology's Promise,* William Halal delivers the differentiation key that entrepreneurs have needed. Not since *Megatrends* have we been given such a clear view into the future of technology and its impact on humanity. Thanks, Bill." – **Jim Blasingame,** *small business expert, author, and columnist, award-winning radio host,* The Small Business Advocate Show

"Fifteen years ago Bill Halal started a project now known as TechCast to forecast the times at which new, emerging, or anticipated technologies would become significant. The first report was well received and in just a few years became an eagerly anticipated annual event. For those new to the fields discussed, the forecasted dates were an eye-opener. For others making their own forecasts, if they agreed it boosted one's self-confidence. If the forecast was significantly sooner or later than you expected, you were forced to reconsider your assumptions about the future. That re-examination of assumptions is the single most important thing one can do in coming to grips with the future. Bill's book develops in engaging detail how the process works, the forces driving the changes implicit in the forecast, and the likely outcome of those changes." – **Joseph Coates,** *Futurist, Washington, DC*

"Bill Halal is a pioneering contributor to the field of science and technology forecasting, especially applied to broader societal issues. He has advanced both the level of technique and the level of wisdom available for more far-ranging and more profound analysis. His work is both a challenge and an encouragement to those of us who believe that it is possible to develop policy for the future on the basis of systematic assessment. It also strongly underscores the now central role of science and technology as drivers in the world civilization that is struggling to emerge." – **Leon Fuerth,** *Research Professor of International Affairs, Elliott School of International Affairs, George Washington University, Washington, DC*

"In this wonderful new book, Bill Halal gives us an image of a positive future based on advances in AI, green technology and the development of a mature global governance system. While the thrust of his theory is technology focused, Halal takes a broad view of technology seeing biology as a technology of life and spirituality as technologies of consciousness. This book is excellent reading and inspiring." – **Professor Sohail Inayatullah,** *Tamkang University, Taiwan, the University of the Sunshine Coast, Australia*

"Bill Halal's personal grasp of emerging technologies coupled with the excellent predictions of the TechCast Panel makes the job of a planner far easier. *Technology's Promise* is a terrific and dynamic roadmap to the future." – **Mike Jackson,** *Chairman, Shaping Tomorrow, London*

"*Technology's Promise* is of great value to those in business and government who are active in forecasting and managing technology advancement." – **Zhang Jingan,** *Chief of Science and Technology, Daily Vice Minister, Ministry of Science and Technology, People's Republic of China*

"Through millions of years of striving to be selected for fitness to survive, the crowning achievement of the human mind clearly is the exponential growth in Knowledge and its applications. Modern humans started out pursuing an understanding of the external world, and since the latter half of the 20th century, we turned inwards and are now beginning to understand ourselves and the microcosm. As distinguished Professor Halal correctly points out, we stand as a species on the brink of boons and banes caused by our technological prowess. Will we justify the salutation of 'homo sapien' or will we fall prey to the inevitability of extinction so strongly foretold in the historical record of evolution? Read Prof. Halal's book. He answers the question brilliantly." – **Walter Kistler** and **Sesh Velamoor,** *Foundation for the Future, Seattle*

"I was deeply impressed by this remarkable work. Prof. Halal's bold method of forecasting has given us a constructive new way to think about how the Technology Revolution is likely to change the world. We are pondering the implications for South Korea carefully and deeply." – **Young-Tak Lee,** *Chairman, Korean Stock Exchange, Seoul*

"Leaders in business, government and universities need foresight. They take big bets on the future. This book will be an extremely valuable aid to their thinking." – **Michael Maccoby,** *PhD, President, The Maccoby Group, Washington, DC and author,* The Leaders We Need, and What Makes Us Follow

"Bill Halal has dedicated his career to tracking the implications of new technologies. This book is especially fascinating on the social and economic impacts of technological change in manufacturing, robotics, and healthcare; as well as the globalization of commerce, communications and security issues; and the impact all will have on income gaps and global prosperity. A thorough discussion on the critical technology issues facing humans in the 21st Century." – **Tim Mack,** *President, World Future Society, Washington, DC*

"I have for years admired Bill Halal's skillful and determined work in developing the TechCast system. It gathers expert views of the future all around the world in a systematic and inspiring way. Now Bill has done a great synthesis and developed scenarios, which go far beyond what normally is understood with the future of technologies. *Technology's Promise* is a must for any futurist or a person interested in our global future." – **Mika Mannermaa,** *Dr of Economics, Managing Director, Futures Studies Mannermaa Ltd, Helsinki, Finland*

"What Professor Halal has done is truly amazing. First, the methodology defies history; you are not supposed to be able to predict technological innovation. Second, he has represented this clever process and its outcomes in an incredibly readable book. The reader's eyes pop in and out on discovering the pace and space of technological breakthroughs. For those who get paid to be concerned about technology and its implications – and for those who don't but still care – if you buy only one book this year, this is it." – **Dennis McBride,** *President, Potomac Institute for Policy Studies, Washington, DC*

"It is likely that Bill Halal has thought more – and more systematically – about the future of technology than anyone you know. For 15 years, his TechCast program at George Washington University has been tracking the likely emergence of technologies that will change your life and the planet. Guaranteed. Here's the big picture in one, easy-to-read place!" – **John L. Petersen,** *President and Founder, The Arlington Institute, Berkeley Springs, Virginia*

"What were once separate scientific disciplines now work together to greatly accelerate technological progress. IT accelerates biotech which accelerates nanotech, which accelerates IT, etc. Bill Halal's method of synthesizing the viewpoints of 100 specialized technology futurists captures the surprising trajectory of this emerging critical mass of technological progress. T*echnology's Promise* is a deep, readable and compelling look into our most likely futures." – **Gifford Pinchot III,** *Co-founder & President, Bainbridge Graduate Institute, Seattle*

"Navigating through game changing technologies is critical to an organization's strategy. *Technology's Promise* provides enterprise leaders with a programmatic view into the future technology landscape, enabling superior strategic decisions." – **San Retna,** *Vice President, Program Management Office, Safeway, and Chair, Enterprise Portfolio Management Council*

"Bill Halal has written an indispensable guide to the future for entrepreneurs, whom we rely on not only for the growth of our economy but also to benefit from new technologies in the pipeline. Bill's unique access to expert information about what's coming next also provides a powerful tool for strategic planners and thinkers." – **John W. Rollins,** *Serial IT Entrepreneur and Professor of Entrepreneurship, Washington, DC*

"*Technology's Promise* offers a vision of a unified world emerging from a virtuous cycle of knowledge. Based on expert opinion on the scientific and technological revolution, the book provides an antidote to doom and gloom; not by ignoring the problems that beset us but by demonstrating that we are creating the science, the technology and the tools to remake society and ourselves to reach a new level of civilisation. This book is a primer for change. Read it, absorb and act!" – **Harry Rothman,** *Editor,* Technology Analysis & Strategic Management, *England*

"Halal offers a breathtaking and immensely practical view of what the future holds for us all. *Technology's Promise* is a guide and an inspiration for action in the face of the immense changes that lie ahead." – **Paul Saffo**, *author and forecaster, Institute for the Future, Silicon Valley*

"Thriving with the accelerating pace of technological change has become a defining personality trait for the 21st century. Now more than ever the regular consumption of high-quality future studies is needed to build foresight consciousness. This passionate, enlightened book with its superlative expert forecasting will help prime you for the staggering opportunities and demanding challenges of our times. Read it and play your own vital part in the transformation of global consciousness now underway. It will help you appreciate both our extraordinary present and the unprecedented future ahead." – **John Smart**, *President, Acceleration Studies Foundation, Los Angeles*

"The 21st century is one of rapid change and globalization. Those who can create and predict the future, as well as operate in it with confidence and knowledge, are the winners. Professor Halal's work, *Technology's Promise*, is a ***must read*** for those who aspire to be winners. Few scholars and practitioners have access to both a field of oracles as well as research and insights from a life dedicated to this subject. Professor Halal has successfully integrated the ingredients of art and science that are the DNA for success in the new world. Read it, and be a winner." – **Michael Stankosky**, *DSc, Professor of Engineering Management & Systems Engineering, Co-director, Institute for Knowledge & Innovation, George Washington University, Washington, DC*

"A fascinating analysis on how the Technology Revolution will transform business and society. Prof. Halal's Virtual Trip Through Time challenges our views of the future. This book can be extremely useful in adapting to the changes we are facing." – **Dr Sylviane Toporkoff**, *Professor, University of Paris, President and Founder, Global Forum/ Shaping the Future*

"Understanding the future of technology and its impact now has become a must for any critical planning. *Technology's Promise* does not come out of Dr Halal's speculation but is based on more than one hundred diversified experts' collective intelligence. Providing solid, evidence-based implications, the book really offers readers more persuasive trends of technology and its effects on social organization. If you are looking for an engine of collective intelligence on technology, this book and its engine www.techcast.org provide solid direc-tion." – **Dr Rong-I Wu,** *Chairman, Taiwan Stock Exchange Corporation Former Vice Premier of the Executive Yuan of Taiwan*

TECHNOLOGY'S PROMISE

EXPERT KNOWLEDGE ON THE TRANSFORMATION OF BUSINESS AND SOCIETY

William E. Halal

palgrave
macmillan

First published 2008 by
PALGRAVE MACMILLAN
Houndmills, Basingstoke, Hampshire RG21 6XS and
175 Fifth Avenue, New York, N.Y. 10010
Companies and representatives throughout the world

PALGRAVE MACMILLAN is the global academic imprint of the Palgrave Macmillan division of St. Martin's Press, LLC and of Palgrave Macmillan Ltd. Macmillan® is a registered trademark in the United States, United Kingdom and other countries. Palgrave is a registered trademark in the European Union and other countries.

ISBN-13: 978–0–230–01954–6
ISBN-10: 0–230–01954–4

This book is printed on paper suitable for recycling and made from fully managed and sustained forest sources. Logging, pulping and manufacturing processes are expected to conform to the environmental regulations of the country of origin.

A catalogue record for this book is available from the British Library.

A catalogue record for this book is available from the Library of Congress.

10 9 8 7 6 5 4 3 2 1
17 16 15 14 13 12 11 10 09 08

Printed and bound in China

Dedication

To the creative scientists, engineers, and entrepreneurs who bring mere possibilities into reality. May this book help guide the path ahead.

Contents

Acknowledgements

I am grateful for the support of my publisher and staff at Palgrave Macmillan for producing and distributing a fine book. Many others deserve recognition. Owen Davies provided keen editorial skill in helping me shape my writing into a more inviting form. I want to acknowledge the judgment of colleagues who reviewed the manuscript and offered insightful suggestions at various stages of development: Richard Donnelly, Chairman of the Department of Information Systems and Technology Management at George Washington University (GWU); Edward Cornish, Founder of the World Future Society; Michael Marien, Editor of Future Survey; and Tom Lombardo, Professor of Psychology at Rio Salado College. Scholars owe much of what they learn to the stimulation of students. Mine at GWU patiently listened to half-baked ideas while I searched for understanding. A particular student, Evan Faber, should be recognized for helping with the graphics. My former PhD student, now Dr. Ann Wang, provided continued support. I was moved to see so many of my TechCast experts offer useful guidance: Denis Belaguer, Dennis Bushnell, Jose Cordeiro, Adam Gerber, Aharon Hauptman, Henry Heilbrunn, Lester Ingber, Michael Jackson, Peter King, Jeff Krukin, Al Leedahl, Xin-Wu Lin, Robert Locascio, David Luckey, Michael Mainelli, Mika Mannermaa, Amy Oberg, David Passig, John Sagi, Yair Sharan, Art Shostak, and Richard Varey. A total of 100 experts listed in the Appendix serve on our Expert Panel doing the hard work of creating forecasts. And the folks at Unifusion brought professionalism to the latest version of our website www.TechCast.org. To all these colleagues, partners, and friends, I express my deep gratitude for their role in producing *Technology's Promise*.

Foreword: Basic Training for the 21st Century

Do you worry that your job may someday be replaced by a machine? Debate whether to buy a new car now or hope that the safety and convenience of a high-end Mercedes will find their way into more affordable models before your aging Ford turns to rust? Wonder how you will cope with tomorrow's consumer electronics when you can't program today's DVD player? If so, you have picked up the right book.

If not, keep reading anyway. Bill Halal has a lot to tell you about the future of technology and how it will affect our lives. These are things you need to know.

Even if you do not worry about the effects of technology – even if "techie stuff" doesn't interest you at all – you can feel the world speeding up around you. This is not an illusion. The pace of change really is accelerating.

Back before World War I, when "high technology" meant chemistry, it took about 40 years for the discoveries of basic science to find their way into daily use. By the 1950s, this product cycle had shrunk to only 30 years. The miracle plastics that Baby Boomers saw on television when they were children grew out of basic discoveries made in the 1920s and '30s – about when the first televisions were invented.

Electronics really broke open the floodgates. The first practical transistor was invented in 1947. The first integrated circuits were patented in 1959 and 1961, but it was roughly ten years later before microchips were cheap enough to use in consumer products. Pocket calculators reached the stores in 1971, and the first personal computer – the IMSAI 8080 – was in production by December 1975.

Today, if a personal computer can be found in a store, it is already obsolete. The next two generations of PC are already in the labs, waiting to enter mass production. The product cycle, from basic discovery to mass marketing and replacement with an even better model, has shrunk from 40 years to six months or less.

So it is with most consumer products. If consumers welcome a new item, knock-offs will be for sale on eBay within a month and cheaper, more powerful models will flood the market three months later. Look at mp3 players for just one example. It takes something really special – like the iPod's eye-catching design and inspired convenience – to hold its market share against the competition.

> "This book provides basic material for planning a secure, productive life in an age of constant change. It is a 'heads up' that no one but Bill Halal could have given."

Now that the 20th century is well behind us, this trend is spreading into medicine, materials, and many other fields. Eighty percent of the scientists, engineers, technicians, and physicians who ever lived are alive today, and they are trading ideas in real time on the Internet. No wonder things are changing so fast! These technologists and their heirs will continue to shake our world when today's young children reach old age – if by then medicine has not made old age obsolete.

In a world of pandemic change, it is all too easy to be caught unprepared. We easily imagine that our jobs are secure or that the economy will remain prosperous and stable for a few more years. Then Chinese competitors bring out a product that is as good but costs half as much – Detroit, you have been warned! – or turmoil in the Middle East sends the price of oil through the roof and real estate prices fall through the floor. The only way you can afford not to look ahead is if you have already retired with enough money to outlast any conceivable misfortune. For the rest of us, forecasting is a basic part of life – rather like doing our taxes – whether we know it or not. This is where Prof. Halal can help.

There are many ways to anticipate the future. You can collect mountains of data, weigh it carefully, stare off into space for a few moments, and make a wild guess, as many economists seem to do. You can try to identify trends in global affairs and project where they will lead. (This is what we generally do at Forecasting International.) Or you can use what almost certainly is the oldest technique of all: ask someone you believe to be an expert. People have been doing that at least since the classical Greeks consulted the oracle at Delphi, some 2,800 years ago.

At George Washington University, Prof. Halal has modernized the ask-an-oracle method of forecasting. Instead of consulting one seer about the future of technology, he periodically questions roughly 100 of them. And instead of being general-purpose priestesses, Bill's oracles are experts in their fields, from medicine to computer technology and rocket science.

What this process buys you is confidence. If one supposed expert tells you that in twenty years we will all be flying around the world on woven-titanium carpets, you may just have picked the wrong person to ask. But if several people with solid records in their field all come up with reasonably similar predictions, give or take a few years, you can reasonably believe that they have given you an accurate look into the future.

That is pretty much how it has worked for Bill Halal's TechCast Project, which was born as a college program and has now grown into an independent research project. Since its founding in 1985, TechCast has produced some of the most interesting and accurate technology forecasts now available. At Forecasting International, we have relied for years on TechCast's annual summary prediction of future technology. Other clients who have recognized the value of TechCast's work include the Asian Development Bank; the U.S. Environmental Protection Agency, Food and Drug Administration, National Aeronautics and Space Administration, and Department of Defense; as well as private corporations such as AT&T, General Motors, and IBM.

You are likely to find TechCast's predictions interesting and valuable as well. In the pages ahead, you will find wonders to match those of any old-fashioned carnival sideshow. They had sword swallowers and bearded ladies. The future Bill Halal presents has robots that can clean your house, drive you to work, or float through your blood vessels killing cancer cells before they form tumors than could kill you. It has computers as powerful as the human brain and capable of reading our thoughts. There are elevators to space and vacations in orbit. There are artificial organs and medical treatments tailored to our DNA and the promise of healthy, productive lives far longer than our natural span. Yes, there are wonders here to amaze and entertain.

After the show, the Wild Man from Borneo took off his costume, changed into street clothes, and went out for pizza with the no-longer-bearded lady. But the wonders in this book are all real. They just haven't arrived quite yet.

When they do, our lives will change again and our expectations will be obsolete – unless we have looked far enough ahead to see where we are going. And it is here that Prof. Halal makes what could be his greatest contribution. Not only does he tell us what is coming, he analyzes how all these technological changes will alter our world and our lives. Even if the future does not exactly match his scenario – and it won't; no forecast ever survives contact with tomorrow's experience

– these final chapters in the book give us a useful framework for understanding. As we see how the future deviates from this scenario, it will be easier to adjust our own expectations.

No one with any sense drives down a highway without looking at the road ahead. But that is roughly what we are trying to do when we make plans without solid, reliable information about the future of technology, the most powerful force now changing the world. This book provides basic material for planning a secure, productive life in an age of constant change. It is a "heads up" that, among all the forecasters we know, no one but Bill Halal could have given.

If you are a techie, read the pages ahead for pleasure and accept the information as a bonus. If you spend your days making hand-thrown pottery and loathe technology more advanced than your kiln, think of this book as the spinach in your information diet. Either way, read on. This is basic training for the rest of the century.

Dr. Marvin J. Cetron
President, *Forecasting International*

Dr. Marvin Cetron is one of the preeminent forecaster-futurists in the world. For some 50 years, he has pioneered corporate and government forecasting, developing many of the techniques that other forecasters now use daily. He remains one of the most active and respected practitioners in this field. He has consulted for roughly 450 of the Fortune 500 firms, 150 professional organizations, and 100 government agencies. He served as an advisor to the White House for every administration, Republican and Democratic, from the time of John Kennedy through the Clinton years.

Preface: Discovering the Forces of Transformation

The relentless power of technology can be seen in a small, everyday product that receives little attention. Travel anywhere today, and you will find that almost all tourists now use digital cameras. As Liquid Crystal Display (LCD) technology and digital storage capacity improved, we reached the critical point a few years ago when digital photography became feasible, and soon the entire industry of film-based cameras was overturned by simply moving around digital bits instead. Nikon, once a mighty force in film photography, now sells only one model of 35mm camera. Eastman Kodak dominated the industry for decades, but has been struggling to adjust for years, laying off tens of thousands of employees and replacing its entire line of products and services. In terms of social impact, the digital camera unleashed floods of photos and video that now populate the new participative Web 2.0 – MySpace, YouTube, blogs, wikis, and more sure to come.

The same revolutionary force is at work as biotech extends lives, alternative energy replaces oil, robots and artificial intelligence (AI) take over human jobs, and endless other innovations reach the take-off point. As the forecasts presented here will illustrate, today's mounting knowledge in all technical fields seem destined to transform society during the next 20 to 30 years. We had best prepare for the coming upheaval.

Technology's Promise provides a roadmap of the terrain ahead. Based on a program of research that summarizes the available knowledge and the judgment of an international panel of 100 experts, this book presents comprehensive forecasts of breakthroughs in energy and the environment, manufacturing, robotics, information technology (IT), e-commerce, medicine and biogenetics, transportation, and space.

We then draw on decades of social studies to understand the institutional barriers to technological change and the pivotal concepts emerging to transform business, government, and other institutions for a knowledge-based global economy.

Because AI is challenging human intelligence, I also examine the difference between AI and human consciousness and survey the many techniques being used to shape consciousness.

We conclude by integrating all forecasts into longitudinal scenarios that outline the most likely path ahead using "macro-forecasting." The most prominent feature of this analysis is that a "global crisis of maturity" seems likely about 2020 to 2030. It may be the most important event of our time.

The book's goal is to provide a compelling new perspective that helps make sense of today's seemingly insurmountable challenges. It can be used as a framework to guide your organization and personal life more wisely.

ORIGINS

Books are like babies in important ways. Apart from the "hard labor" involved in producing both, their character is largely inherited from different sources. *Technology's Promise* is the product of two enduring streams of study that came together during a long gestation to produce a lively intellectual offspring.

I was educated as an aerospace engineer, served as an Air Force officer, worked on the Apollo Program and in Silicon Valley, so I've always loved science and technology. As a boy I was fascinated by taking things apart and figuring out how they worked. I loved the logic of physics, marveled at flight, and pondered electricity.

However, I also felt the technical world was isolated from the "real world" of commerce, politics, and society. People were sure I'd gone mad when I left the Apollo Program to get an MBA and PhD in economics and the social sciences. Then I taught business, strategy, and other "soft" fields for many years while trying to understand social systems. But this only left a longing for the precision and hard reality of science and technology, which led to the work reported here.

Technology's Promise integrates the hard world of technology and the soft world of social science into a more comprehensible whole. I think I can now bring a richer perspective to this sixth book of mine, showing not only how the technology revolution is likely to unfold, but also how it promises to alter social institutions, the global order, and even what it means to be a modern human.

This preface offers a short intellectual history of *Technology's Promise*. Here's the story of how this work developed, written from the "inside out" to identify the valuable lessons learned over the past 30 years – the power of massive technological change, overcoming institutional obstacles that block its path, the rising role of human consciousness, and the realization that the entire world is evolving through its own far greater life cycle to form a coherent globe. These crucial insights form the conceptual foundation of the chapters that follow.

LEARNING TO FORECAST EMERGING TECHNOLOGIES

This technological perspective struck me forcibly a few decades ago. Having experienced the end of the Great Depression, World War II, the Eisenhower Boom of the '50s, the student revolt/civil rights movement of the '60s, the collapse of Communism, invention of the PC, and recently, the Internet, I needed a way to understand this trajectory in which life was being played out. The evolution of technology explained so much, and I find it endlessly useful in guiding my study of how the world is changing. I became so absorbed that I spent 20 years developing a theory of technological evolution. It was embarrassing to tell people I was working on it so long, but I published three different versions at different stages of progress.[1]

One of the biggest challenges is finding a way to study technological change. Good technology forecasting is rare, especially on a systematic basis with periodic updates. Most academic research is densely written, and university pressures to publish encourage work that tends to be trivial. Possibly the biggest obstacle is that the social sciences, beset by physics envy, are so focused on statistical analyses of large data sets that they unwittingly ignore the biggest issues confronting society. Countless professors across the land are busy performing multiple regression analyses of theoretical models using dozens of variables, hardly noticing that corporations, governments, and the other subjects of their scrutiny are being transformed daily into unrecognizable new forms. Michael Crow, President of Arizona State University, says he is fighting academic traditions "linked to the past," which have produced a nation so focused on narrow interests that it's hard to accomplish much.[2]

The leading edge in social research today is moving online where more flexible and powerful tools are available. I began to develop methods of this type with my first book, *The New Capitalism*,[3] and have now refined it into a reliable and powerful approach to anticipating technological breakthroughs. The elements of this method are described in the introductory chapter – scanning, analysis, survey, action, feedback – and comprise a new epistemology of "online research/learning systems" increasingly being used in almost all settings.[4]

When the introduction of PCs started a swelling wave of innovation, I began to forecast critical breakthroughs using a mail survey of experts. It was crude, but the response to our first publications was striking as a flood of email, letters, and phone calls told me there was a huge need for better forecasts. As a university scholar, I always felt that publishing academic work was very similar to throwing it into a hole, never to be heard from again. So this caught my attention. The method was improved every two years or so, and has now become a sophisticated website, www.TechCast.org, that may provide the most comprehensive forecasts available.

I've also come to see the need to integrate studies and data with feelings and inspiration, so I also use a healthy dollop of scholarly judgment, personal insight, and just plain intuition. These forms of "soft knowledge" can mislead if not tempered by facts, yet experience shows sound forecasts require careful judgment. Inspiration without evidence may be ideology, but data without meaning is purposeless.

What can we call this approach? I am not sure myself as it contains elements of different paths to understanding. It is about the future, but it's more than that. TechCast is a science-based form of forecasting designed to guide the technology revolution. Because all strategy hinges on adapting to a changing world, this also forms the basis of strategic planning, product development, and organizational change. Technology is the foundation of society itself, so this perspective helps understand how social structures are being transformed and it unravels daunting problems like globalization, the energy and environment crisis, and other controversial issues of our time.

Now that academics whose work is guided by evolution call themselves "evolutionary biologists," "evolutionary anthropologists," etc., perhaps this approach can best be described as "evolutionary futurism/forecasting/strategy/social science."

CUTTING THROUGH ORGANIZATIONAL GRIDLOCK

This work also led to the realization that the present social world we are comfortable with is rather arbitrary, dysfunctional in many ways, and merely a temporary state of affairs. Most of us view life within the limits of the present because it's so obvious, just as it was once obvious the world is flat. This limited temporal perspective fails to recognize that the world we inhabit is but one small step in a long and constant process of transformation. Just a few short decades ago we were perfectly content using slide rules and calculating machines, typing pools, and a few major TV networks. Women worked at home, and men retired after 40 years with the same employer.

While the technology revolution is promising, it faces enormous obstacles posed by today's outmoded social systems. Many have noted that business, government, education, religion, and all other institutions seem frozen in a sort of "organizational gridlock" that blocks change. I was stunned early in my career to hear a senior manager explain that the only goal of business is to maximize profits. Thinking as a rational engineer and scientist, it struck me as terribly misguided to ignore the role of employees, clients, suppliers, the public, and other parties who were clearly crucial to economic success. A host of similar problems persist though they defy logic because people are convinced that institutions are inviolate and change is impossible.

After many years studying progressive practices and illustrating how they could bring rationality to organizations, it dawned on me that most people simply are unable to consider the big changes that are needed. At one point, I used a slide called "Bullshit Bingo" containing a bingo scorecard with entries of common buzzwords – "Win-Win Solution," "Think Outside the Box," "Alignment," etc. – to be checked off in meetings until one can shout "Bingo." It's dumb, but it always drew guffaws, making the point that most people do not take present approaches to organizational change seriously.

This led to the insight that the underlying obstacles are "institutional" rather than "organizational." As we will see in Chapter 8, most organizations can handle the routine stuff fairly well – teamwork, process design, leadership, etc. They are frozen in time, however, by their inability to reconsider the underlying values that are commonly accepted as dogma – especially the dominant role of power and money. I have come to appreciate the enormous resistance to altering such strongly-held beliefs, and to see that decades of discussion, experimentation, legislation, and crises are needed.

It may take time, but various trends suggest that one of the great impacts of the technological revolution should be to redefine institutions for a knowledge-based world. Chapter 8 will outline the results of original research and countless examples showing how institutional norms are yielding to two major themes:

Driven by the demands of raging complexity and turbulence, progressive corporations increasingly use "self-managed teams" that form entire "self-organizing systems" better able to use knowledge at the grassroots. Hierarchical control is thereby replaced by small "internal enterprises" acting from the bottom up.

And with the public increasingly demanding social responsiveness, corporate executives are broadening their domain to work with employees, customers, and the public as collaborative partners. All interests can thereby be served by forming a quasi-democratic "corporate community," including earning more money for stockholders.

I think these two central themes offer the prospect of a new form of political economy that carries Western principles of free enterprise and democracy to their logical conclusion. We will see in Chapter 8 how this synthesis of conservative and liberal values could reconcile the deadlock between right and left that now paralyzes the U.S. and much of the world. Unifying these opposing views could also resolve nagging old dilemmas, like the social resistance to corporations, the unsustainable U.S. healthcare system, and possibly terrorism.

CONSCIOUSNESS AND THE CRISIS OF MATURITY

The power of information was celebrated when the field of "knowledge management" rose to the top concern of corporate strategy a few years ago. It was marvelous to see an invisible resource like knowledge gain such prominence.

After a flourish of progress, I realized that an even more powerful domain of "consciousness" lay beyond knowledge, and that is where the real action takes place. Interest in the study of consciousness is exploding now as computers model the brain, but I think scientists are going to be disappointed when it becomes clear that human consciousness transcends information and knowledge altogether. Chapter 9 explores how the limits of formal knowledge may force us to accept the reality of human awareness as a distinctive and powerful force governing life.

Our most dramatic insight is that we are likely to witness a culmination to this evolutionary process in two to three decades. As noted, the last chapter integrates all our forecast data over time into four scenarios outlining the likely path of the technology revolution. What really stands out is that the forces of globalization almost inevitably lead to a global crisis of maturity. This crucial transition is defined by the following two sides of all crises – threat and opportunity.

Both evidence and knowledgeable opinion confirm we are moving to a more populous world that is largely industrialized and intelligent, but that also poses unprecedented risks of environmental damage, energy shortages, climate change, weapons of mass destruction, and other threats that require sophisticated responses unimaginable by present standards. These crises seem insurmountable because the present world system is not sustainable, and a shift to a more sophisticated system is unavoidable to avert disaster. As former President Bill Clinton once observed, "There is no world system." We need one urgently.

Some of these dilemmas may be ameliorated by advances in economics and technology, but they can only be resolved through the logic of evolution, which is now moving *beyond knowledge*. At about 2020, the very time when the planet is likely to teeter between calamity and salvation, routine human thought should be automated by far more sophisticated IT networks, a second generation of more powerful computers, and good AI. As various forms of machine intelligence take over common mental tasks, we will simply move up another level on the evolutionary hierarchy to address these global challenges.

It's impossible to really grasp the reality of a different era, but I think an "age of consciousness" is likely to emerge, focusing on higher level understanding and on working out together the tough existential choices needed to survive. Chapter 9 will show how strategic dialogue, collaborative problem-solving, conflict resolution, ceremonies, mediation, prayer, and other yet unknown "technologies of consciousness" may offer the next logical step in this evolutionary process.

This looming crisis of maturity may not prove catastrophic if acted on in time, but there seems to be an unavoidable need for a major turning point as the problems of industrialization, energy, environment, conflict, etc. reach critical levels about 2020 to 2030. The obstacles are enormous, of course, although it is precisely because so many people are so deeply concerned that a major shift in global consciousness is underway. Yes, I know this seems utopian, but read on, and I think you will find the argument thought provoking, if not convincing. I was fortunate to know the late Admiral Arthur Cebrowski, who pioneered the U.S. military's Office of Force Transformation. While working on an article together, I was struck by the refreshing candor of his views: "We have to recognize that a major transformation is inevitable."[5]

At a time of revolutionary scientific and technical mastery, it is ironic that the central role played by consciousness in this drama of the high-tech future leads back to the teachings of ancient prophets, who all saw that a domain of "spirit" pervades life. Moses, Christ, Mohammed, Buddha, and other founding religious leaders may have missed a lot, but they all pointed toward this great unseen force that puzzles us still. Our challenge is to understand the mystery of consciousness in modern scientific terms, and use that understanding to shape awareness. The fate of the world hinges on our ability to harness the human spirit so we may guide ourselves more wisely.

William E. Halal
November, 2007
Washington, DC

Notes

1 The most recent version is Halal, "The Life Cycle of Evolution: A Macro-Technological Theory of Civilization's Progress," *Journal of Futures Studies* (August 2004) 9:1, pp. 59–74.
2 David Ignatius, "Solving Stovepipe America," *Washington Post* (June 7, 2007).
3 Halal, *The New Capitalism: Business and Society in the Information Age* (NY: Wiley, 1986).
4 The U.S. Patent Office, NASA, FBI, and most corporations are implementing various forms of Web 2.0 and Gartner expects the field to dominate business soon. See www.TechCast.org for more.
5 Arthur K. Cebrowski, "Seven Secrets of Transformation," in Halal (ed.), *Institutional Change*, a special issue of *On the Horizon* (2005) Vol. 13, No. 1.

CHAPTER 1

Introduction: Guiding Technology's Promise

In the 1960s when today's explosion of high technology first began, we were wowed by the "ape" scene in the classic movie, *2001: Space Odyssey*. The scene dramatically captures the fundamental power of technology to drive human progress, a power we are rediscovering daily. It is one of the great icons of the Information Age.

In the movie, an ape discovers that an old bone makes a great club. For the first time ever, one of our ancestors has created a tool. He can use it to break things. Defend himself. Kill his enemy. After affirming this novel idea, the ape raises his club to deliver a blow previously well beyond his power, and the swelling music announces that a crucial watershed has been reached. Technology has arrived, initiating the upward march of human history. During the next few million years, this pivotal force allowed mere apes to make the long journey to our present world of biogenetic engineering, quantum computers, and space exploration.[1] That is the power of technology.

We know that technology flows out of creativity, knowledge, and inspiration, which may explain the significance of the black monolith in *Space Odyssey*. This second great icon points to another major theme in *Technology's Promise* – the rising role of consciousness. I don't know what the author of the movie, Arthur C. Clarke, had in mind, but the monolith strikes me as the quintessential "black box" containing vast powers shrouded in mystery. It is a fitting metaphor for consciousness.

Within the last decade or so, a mere nanosecond in evolution, an even greater barrier has been breached as sophisticated information systems begin to close this evolutionary cycle. Today knowledge – the very heart of scientific progress – is being harnessed on a massive scale.[2] The decoding of the human genome, for instance, was only possible using supercomputers to decipher the 3 billion bits of information in DNA.

This historic step can be understood as a "virtuous cycle" of continually increasing scientific knowledge driven by the Information Technology (IT) Revolution. Figure 1.1 illustrates how IT improves

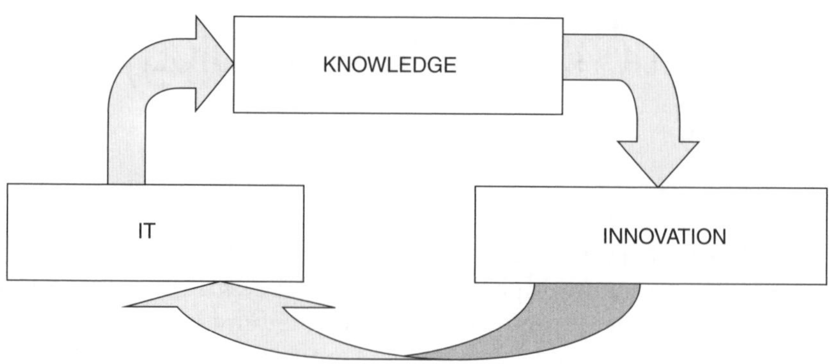

Figure 1.1 The Virtuous Cycle of Knowledge

our ability to acquire knowledge → which then allows more widespread commercial innovation → which in turn improves information systems again → on and on in a spiral of transformation.

This is not simply another scientific advance, but a breakthrough in the process of science and industry itself. Scientific research and commercial innovation are growing in power and speed as the ability to amass knowledge drives technological progress as never before. Some scientists, like Ray Kurzweil, think of it as a "singularity" in which the pace of technological change leaps dramatically during the next 20 to 30 years.[3] Although many think this will cause computer power to surpass human abilities, our discussion of consciousness in Chapter 9 will show that this is likely to prove illusory.

The virtuous cycle of knowledge is opening up dramatic new fields of discovery that are interwoven to build on each other. A central breakthrough is the way the IT Revolution transforms everything – science, economics, culture, and even awareness. We are also discovering that information lies at the heart of life, as exemplified in the way the DNA molecule encodes the information defining each organism. Closely related is the increasing ability of nanotechnology to bond atoms in ways that form a variety of tiny structures, and the role of observation (information again) in governing the unusual behavior of atomic particles in quantum physics.

The emerging scientific perspective, then, sees the world as an intimate interplay between atomic matter, biological life, and information. Some call it the "molecular" age, because it focuses on how information processes organize matter and living organisms at the fundamental scale of the molecule.[4] Many contend that this power to control matter and information is roughly akin to playing God. Craig Venter, the scientist who helped decode the human genome, is now creating synthetic life in his lab.

The growing power of science explains why we are seeing breakthroughs everywhere. As the evidence presented in this book will show, we can now realistically envision renewable energy replacing oil, medical control over the genetic process of life, computer power becoming cheap and infinite, mobile communications at lightening speeds, robots serving as helpers and caregivers, and much more to come. The technology revolution is still at an early stage, and it presents great dangers as well as benefits. But its potential for using knowledge to solve technical problems is so great that it is limited only by imagination and will.

Although the benefits should be vast, we must remember that technology can be easily misused and it often produces disastrous unforeseen consequences. The automobile made modern society possible, but its costs include traffic congestion, air pollution, energy crises, and some 40,000 highway deaths/year in the U.S. alone. If the auto were a new technology that had to be authorized by the U.S. Congress, it is not at all clear that cars would pass muster without severe regulations.

Some think this is a deterministic view, whereas technology is merely an enabler that allows people to make their own decisions. That is certainly true, but it is a serious mistake to downplay the enormous power of technological forces that can sweep away everything in their path. A business owner may dislike IT systems, yet he/she will probably need to use them to survive in a competitive marketplace. We always have freedom, then, but it is constrained by technological imperatives.

The coming upheaval is almost certain to present massive challenges we are not yet equipped for. The world is changing so much and so quickly that most people do not grasp what we are getting into. We lack even the concepts to begin understanding.

THE TECHCAST PROJECT

This revolution in technology will affect all sectors of society, alter the way people work and live, and restructure the world. Little wonder that it is challenging our best minds today. In the pages ahead, I hope to demonstrate that it is possible to understand these changes more insightfully using a carefully organized system of study. For many years I have conducted the TechCast Project at George Washington University and at my own company, TechCast LLC, to forecast emerging technologies.

People everywhere sense the world is passing through great technological change, but they lack convenient, reliable information. TechCast fills this need using a sophisticated website (www.TechCast.org) that scans the literature and surveys 100 high-tech executives, scientists and engineers, academics, consultants, futurists, and other experts around the world to forecast breakthroughs in all fields. (See Appendix for list of experts) No forecast is perfect, of course, but we think this approach provides what I think of as "the best possible answers to tough questions." Results are automatically calculated and distributed over the website to corporations, governments, and others – anywhere in the world, on any prominent technology, in real time.

Our studies show that technological advances, their adoption patterns, and social impacts follow well-defined cycles that can be forecast rather accurately. This is possibly the most complete forecasting system available covering the entire span of technological innovation.[5] Figure 1.2 highlights the TechCast results, showing forecasts for 61 leading technologies organized into seven fields. We will

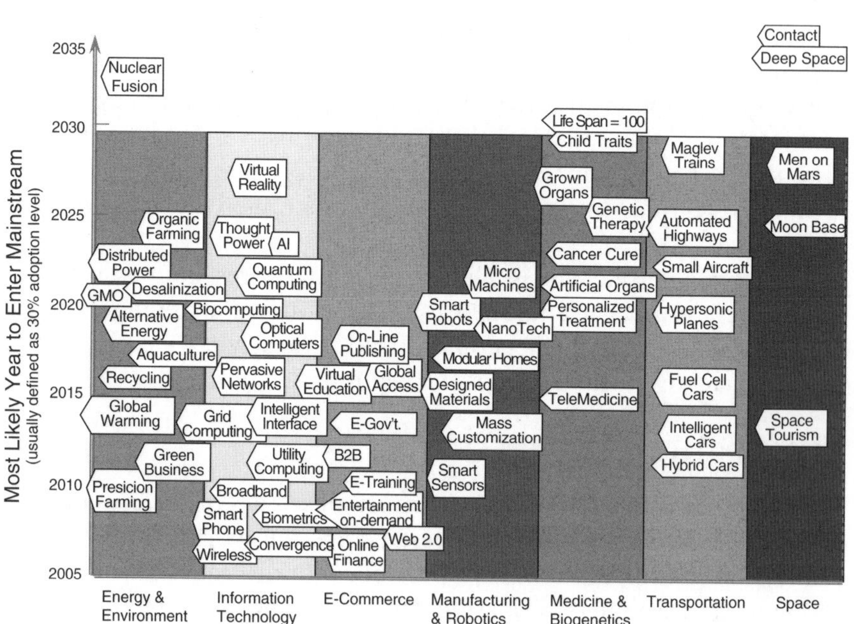

Figure 1.2 Summary of Forecast Results

examine details of these breakthroughs in later chapters, but this overview demonstrates that dramatic advances are underway. They will transform our lives in the years ahead.

The world constantly struggles with huge uncertainty, so how can we justify attempting to forecast such controversial events? There is a long record of distinguished forecasts that proved remarkably accurate. H.G. Wells anticipated a computerized world long before others had an inkling. Jules Verne foretold landing on the moon and atomic submarines, with remarkable accuracy. Arthur C. Clarke defined communications satellites more than a decade before the launch of Sputnik. Peter Drucker described knowledge work in the 1970s. There are many other fine examples, which show that well informed, imaginative judgment can produce remarkably prescient forecasts.

True, we are constantly assailed by predictions that prove naïve. Think back to the common fear in the 1960s that automation would leave us with excessive leisure and nothing to do! We were also told to expect nuclear power "too cheap to meter," a "paperless office," and the stock market to hit 30,000 on the Dow about the year 2000. How can we sort out half-baked ideas from sound forecasts?

The TechCast process is not based on imagination, prophecy, or speculation but on the scientific method. It is empirical in nature, gathering the best background data available and organizing it into a careful analysis of each technology. Experts are taken through these analyses online and instructed to enter their best estimate of when each technology is most likely to enter the mainstream, the potential size of the economic market, and their confidence in the forecast. The experts are not all world-renown, but they represent the leading edge of collective knowledge about technology. To keep the analysis intellectually honest, we make a point of including opposing trends that hinder technology, such as political obstacles, high costs, social resistance, or other barriers. Comments from the experts and new data are used to update the analyses periodically. These forecasts are not snapshots in time but the result of a continual tracking process that improves as technologies "arrive."

Because the TechCast method is basically a system for pooling knowledge, the field of Knowledge Management (KM) offers a useful perspective for understanding the rationale underlying our approach. From the KM view, TechCast is a "learning system" conducted by a "community of practice" to "continually improve" results and approach a "scientific consensus." One of the most vivid experiences of this work is seeing how pooling the tacit knowledge of 100 good minds and cycling through all this information can create forecasts that are remarkably prescient and reasonable.

TechCast has used this method for 15 years on a variety of studies. On average, the forecasts of when a given technology will arrive vary by +/– three years.[6] Some technologies vary widely because they are controversial, while others show little variance because they are well understood. We have recorded arrivals of several technologies roughly within this likely error band of three years. The results are more compelling when considering the fact that the expert panel changed over this time, as did the prospects for various technologies and other conditions. "Prediction markets" have demonstrated remarkable accuracy recently using the same method, but they put teeth into the process by requiring experts to bet real money on their judgments.[7]

This work also holds up well when conducting studies for other purposes. On one consulting assignment, we conducted two parallel studies to forecast energy technologies, one using a group of energy experts and the other using a group of general experts. The forecasts compared almost exactly, usually within one to two years.[8] The method was also used to anticipate the emerging system of business and economics emerging for the Information Age at least a decade before its arrival.[9] In another case, we anticipated a "global information network for science, commerce, and communication" before anyone knew it as the "Internet."[10]

It is often thought that methods like this are subjective, whereas quantitative methods are more precise. However, quantitative methods also involve uncertainty because they require underlying assumptions that often are doubtful. The TechCast method subsumes quantitative forecasts into the background data and allows the judgment of experts to resolve the uncertainty that remains. Experts may have their own bias, naturally, but it is usually distributed normally, washing out in the aggregate results. If the present level of uncertainty is defined as 100%, we have found that this process reduces uncertainty to about 20 to 30%. Our clients say there is nothing else like it.

TechCast can be compared to a stock market. Stock markets basically manage knowledge about capital, and now we need systems for managing knowledge about technology.

A VIRTUAL TRIP THROUGH TIME

The next few pages offer a quick summary of our more intriguing conclusions from each chapter. Think of it as taking a virtual trip through time in order to grasp where we are headed. In Part I, we focus on the conceptual breakthroughs underlying each technological field, the painful dilemmas they provoke, and a simple but compelling story of how each is likely to change the world. Part II addresses the social implications of the Technology Revolution: restructuring business and other institutions, the challenge posed by artificial intelligence (AI) to human intelligence, and four scenarios on a trip through time.

These chapters also examine the critical issues that must be overcome along the way. The mounting crisis in energy, environment, and climate change caused by the spreading of industrialization throughout the globe. The hijacking of IT by terrorists to subvert the system itself. The moral dilemmas posed by the powers of biogenetic engineering. The need to restructure business and other institutions for a knowledge-based world. And the impending shift in human identify as robots and AI take over many of the tasks that have consumed our time.

Each field of technology is summed up in a bubble chart to highlight emerging capabilities that are especially critical, interesting, or strategic. These are breakthroughs with profound scientific implications, big commercial potential, and great social impacts. They can be predicted with high confidence and are immediate enough to take seriously. In other words, they are most likely to affect you, your organization, and your community. All situations are different, but we strive to tease out implications throughout the book to help corporations, governments, and individuals anticipate these profound changes and react with constructive strategies.

In short, the approach is to subsume all available knowledge – driving trends, opposing obstacles, critical issues, and expert forecasts into an accurate strategic assessment. Throughout the text, highlights are drawn from the website to flesh out these ideas, and endnotes offer other sources. References for all these facts and other data are available at www.TechCast.org. Here's a summary of the chapters:

PART I FORECASTS OF THE TECHNOLOGY REVOLUTION

Chapter 2 Transition to a Sustainable World

We start by showing that industrialization is likely to cover most of the globe at about 2030, producing a three- to five-fold leap in the demand for energy and other scarce resources, in pollution levels, global warming, and other aspects of the industrialization-energy-environment crisis. The modernization of China and India alone will double or triple these problems.

Our forecasts show that some of these issues are likely to be resolved over the next 10 to 15 years. Corporations are now moving to green business practices because the inevitability of this transition has made environmental management a competitive advantage, spurring a huge boom in anything green. The issue of global warming is likely to be addressed seriously about 2012, and alternative energy should make a good-sized dent in the use of carbon fuels about 2020.

This chapter concludes that the industrialization-energy-environment crisis actually is a great opportunity in disguise. The transition to a sustainable world will produce an enormous new industry to manage the Earth. It may even serve to unify people after centuries of ethnic, sectarian, and tribal conflict.

Chapter 3 Globalization Goes High-Tech

Chapter 3 will show that the old, smoking factories of the Industrial Age are yielding to intelligent manufacturing systems operating virtually to produce almost anything cheaply, quickly, and customized to order.

Research in materials and nanotechnology is making it possible to design almost any type of product, and mass customization can deliver an endless stream of sophisticated goods customized for each individual. Driven by the logic of cheap labor and new markets, these changes promise to bring material abundance to poor nations over the next few decades, eliminating much of the poverty that blights the planet.

The tension in this emerging world of plenty, however, will be mounting demand for scarce resources like oil, massive loads on the environment, and more clashes between diverse cultures, as in the conflict between the West and Islam.

Chapter 4 Society Moves Online

Advances in broadband, wireless, and AI are inexorably moving life online as computer power becomes cheap, ever-present, and intelligent.

Our forecasts show that today's rapid growth of online entertainment, e-tailing, virtual education, and other such e-commerce services will soon dominate modern economies. Over a longer term, optics, quantum physics, and nanotechnology offer the hope of continuing the gains in computer power when silicon chips are unable to further improve performance (Moore's Law).

Within a decade or so, we could simply speak to high-fidelity images on large wall monitors while working, shopping, learning, and conducting almost all other social functions. You might buy something by simply talking with an onscreen robot that greets you by name, knows all the merchandise and displays it on demand, answers questions, and has infinite patience – the perfect salesperson.

Chapter 5 Mastery Over Life

A variety of breakthroughs in medicine and biogenetics is likely to provide mastery over the process of life itself.

Artificial organs are being developed to replace almost all bodily functions, including parts of the brain, and stem cell research is increasingly able to repair organs. Life extension techniques are expected to raise average life spans to 100 years within a few decades, and possibly beyond the biblical 120 years. Just as the Industrial Age mastered most aspects of the physical world, the Knowledge Age is now making it possible to master the biological world. Yes, it sounds too good to be true, but so did the notion that men could fly, much less travel to the moon.

We also explore how this progress presents social and moral dilemmas that will have to be resolved. How will we make difficult medical choices about stem cell research, designer babies, life extension, euthanasia, and other sensitive matters?

Chapter 6 Faster and Farther

Travel is being reinvented to manage an explosion of global commerce. We will describe the emergence of the "intelligent car," maglev trains floating between major cities on a cushion of air at 400 mph, and Mach 10 hypersonic aircraft that could reduce flying times around the globe from 30 hours to 3 hours.

It may seem that information systems could replace travel, but information forms a virtual world that parallels the physical world. People will always want to visit each other, handle the merchandise, and hammer out tough decisions together. The need for physical contact is inexhaustible, and some studies show that growing virtual contact only makes face-to-face relations more necessary.

The physical and virtual worlds coexist in parallel dimensions, so travel will likely grow alongside the movement of information. Thus, we forecast there will be no rest for the weary road warrior.

Chapter 7 The Final Frontier

Space tourism is likely to become common in one decade, and we are likely to see the establishment of a permanent moon base and a manned landing on Mars in about two decades.

But the ultimate challenge of deep space travel to distant solar systems awaits fundamental breakthroughs in our understanding of physics. The distances of deep space are so enormous and our capabilities so puny that it will take long, intense research to discover ways to traverse them.

Our estimates suggest the needed scientific breakthroughs are likely to arrive about 2050, which coincides with our forecasts for deep space travel. Forecasting anything that far off seems foolhardy, but it is compelling that a variety of sources suggest travel to other star systems is likely about this time.

PART II SOCIAL IMPACTS OF THE TECHNOLOGY REVOLUTION

Chapter 8 Shifting Structures of Society

Here we explore how this technological upheaval is restructuring business, government, medicine, education, and other institutions as a knowledge-based world alters the basis of economics and leadership.

Financial investment powered the Industrial Age when capital was needed to build manufacturing capability. But today speed, agility, knowledge, collaboration, and innovation are the critical factors needed to survive a world of creative destruction, fickle clients, transient workers, and shifting social values. That's why corporations are constantly in flux, scandals like Enron highlight ethical failures, government is struggling to redefine itself, education is going virtual, and rising medical costs are unsustainable.

We will see that two main trends are driving institutional change. Hierarchies are dispersing into "self-managed teams" able to manage complexity by harnessing the knowledge of ordinary people. And the old focus on profit is yielding to a "corporate community" of collaborative partnerships among employees, clients, business alliances, investors, and the public. These two major trends represent a synthesis of the Western ideals of free enterprise and democracy, offering the possibility of resolving the political impasse between right and left that grips the U.S. and much of the world.[11]

Chapter 9 An Age of Consciousness

Here we explore what follows information and knowledge. Chapter 4 shows that the Information Age is likely to mature about 2020, so what's next? Just as the agrarian economies yielded to manufacturing, which is now being eclipsed by services and information, the knowledge economy eventually will run its course as well.

Services have been automating for years (ATMs, airline kiosks, etc.) and wealthy nations now fear that even knowledge – once thought immune to export – is moving off shore to lower-paid people in developing nations. Various forms of AI are spreading, like the intelligent agent that answers your phone calls, smart computers, and cars that talk to us. All-purpose robots are so well developed that the Japanese and Koreans expect to be selling equivalents of R2D2 to families by about 2010.[12]

This "automation of mental work" poses one of the most fascinating issues of our time – *Is there a fundamental difference between machine intelligence and human intelligence?* Despite the fact that about 90% of us are utterly convinced that human thought surpasses sheer information, could we all be wrong? Everybody once accepted the Flat Earth model of the world for millennia. Is science poised for another great revolution demonstrating that we are fundamentally not much more than wet computers? Or will this critical issue force us to accept a domain of consciousness and human spirit as the new frontier? This chapter explores some of the most challenging and fundamental questions now before us.

Chapter 10 Scenarios

We conclude by integrating all forecasts across fields in vivid, decade-by-decade scenarios to explore how this wave of innovation is likely to unfold – a sort of surrogate for time travel.

2010 should see even greater advances in information systems and e-commerce, making most of the world smarter, faster, and fully wired. By 2020 AI will permeate our lives and permit huge advances in telemedicine, virtual education, and e-government. About 2030, industrialization is likely to reach most developing nations, enabling as many as five billion people to live at modern levels. But intercultural conflict, weapons of mass destruction (WMD), and threats of environmental collapse will pose an historic crisis of maturity that challenges basic worldviews. By 2050, this crisis of maturity is likely to be resolved with the emergence of a modernized global society, somewhat like a far larger and more diverse version of the U.S. or E.U. Local wars, ecological disasters, and other troubles will continue but limited to the normal dysfunctions of any social system.

We then move across these scenarios to outline the larger path of civilization's progress. The agriculture, manufacturing, and services stages have largely been completed in modern nations, and we are now passing rapidly through the knowledge phase toward some form of global consciousness. I will show that this movement through ever more powerful stages of development comprises a great "life cycle of evolution," somewhat like the individual human life cycle but vastly larger and longer.[13] Today's crises, and their eventual resolution, can be better understood in the context of this great arc of civilization now approaching the brink of global maturity.

GUIDING TECHNOLOGY'S PROMISE

Whenever I discuss this, I hear great skepticism because people see no way out of today's dilemmas. Friends often advise me not to be "so optimistic." They are afraid it will damage my credibility. However, mere optimism is blind hope without justification. The views presented in this book recognize that the world must mature if it is to survive, and the evidence assembled here presents an entirely plausible path forward. In fact, our forecasts describe the most likely outcome, rather than mere possibilities. They are justified on empirical grounds, rather than wishful fantasy, and they may be overly conservative. There are no guarantees, of course, but they offer a reasonable basis for the type of vision needed to meet today's challenges. Without such aspirations, nothing would change. We would remain captives of today's world.

Everything I have learned tells me that the next stage in social development will focus on building some higher-level system of governance organized around a global community. Within the past two-hundred years, civilization has advanced from farming the soil, to manufacturing goods, to managing global organizations, and now to harnessing knowledge. Some of our individual forecasts may prove wrong, but the broad direction seems to lead beyond knowledge to a new global consciousness, as many others have foreseen.[14] This hardly means the future will be a utopia, as war, economic failures, and political stalling are not about to disappear. But with a modicum of hard work, luck, and heightened awareness, this wave of technology promises to produce a coherent form of global order about 2050, and possibly sooner.

The key to understanding this transition lies in seeing that technological evolution comprises a natural life cycle of the entire planet, much like the life cycle of any organism although infinitely larger. Just as the Industrial Revolution was powered by the need to overcome poverty, and the IT Revolution is crucial for managing complexity, the world is today in the throes of a "mental revolution" to resolve the global crisis of maturity. Think of the similar crisis every teen faces in shedding the baggage of youth to become a responsible adult. Whether a teenager or an entire civilization, the challenge is much the same – grow up or die.

Things look especially bleak today because that's the normal situation facing any system struggling through maturity – a teenager, a nation, or a global civilization. Those who have raised youngsters quickly learn about the mercurial anguish that is normal in such transitions. Our present experiences are rooted in the past, however, so they are not a feasible basis for thinking about the global transition, any more than the experiences of childhood allow us to anticipate adulthood. Nobody plans these passages; they are natural events in the cycle of life.

Today, the rise of a global civilization possesses a life cycle all of its own that is unfolding very rapidly. The lightning-fast spread of information and knowledge, at the very time the world is unifying into a global network of economic systems, has provoked a series of mental shifts to address the challenges ahead. We have accepted women in power, transformed planned economies into free markets, and begun to protect the environment. Now the really tough challenge of shifting consciousness lies ahead. Although the transition to global maturity could reach critical mass any time, the evidence suggests that it is most likely about 2030 when conditions become critical.

People tend to place such prospects in the distant future, if they accept them at all. But the power of IT systems is now hastening advances as never before, accelerating the passage to a coherent world order in our lifetimes. A 50-year-old today is likely to witness the global crisis of maturity at 70 years of age and the emergence of a global order at the age of 90. As Chapter 7 notes, life extension should raise average life spans to 100 years by about 2030, so these are realistic goals. Even an 80-year old today could easily live to see the crisis of maturity. This is not merely an intellectual exercise, therefore, because most of us will witness this highpoint in civilization.

The technology revolution is fraught with risk and clouded with uncertainty. Yet by carefully studying these possibilities, we can better fulfill the great promise held out by technological advancement. The physicist Freeman Dyson called technology "a gift from God." That gift is growing rapidly now and raising very provocative hopes and fears, much like a teenager getting ready to drive off in the family car for the first time. Yes, he could have an accident, but with the guidance of parents, the experience should help him grow into a competent adult. To avoid squandering the enormous gift of technology's promise, I hope this book can help guide it to success.

Notes

1 Halal, "The Life Cycle of Evolution: A Macro-Technological Analysis of Civilization's Progress" *Journal of Future Studies* (August 2004) Vol. 9, No. 1, pp. 59–74.

2 See William E. Halal, *The Infinite Resource: Creating & Leading the Knowledge Enterprise* (San Francisco: Jossey-Bass, 1998).

3 Ray Kurzweil, *The Singularity is Near* (NY: Penguin Books, 2005).

4 Mihail C. Roco and William Sims Bainbridge, *Converging Technologies* (Arlington, Virginia: National Science Foundation and Department of Commerce, 2002).

5 Publications on technology are popular now, but we are not aware of others that provide such comprehensive, authoritative forecasts. Some of the best recent works include David R. Howell, *The Global Technology Revolution 2020* (Santa Monica, CA: RAND Corporation, 2006), The UN Millennium Project conducted by Jerome Glenn and Ted Gordon; and the trends provided by Marvin Cetron of Forecasting International. However, none of these sources contain actual forecasts.

6 Halal, William E. *et al.* The GW Forecast of Emerging Technologies. *Technological Forecasting & Social Change* (1998) Vol. 59, pp. 89–110.

7 Justin Wolfers and Eric Zitezwitz, "Prediction Markets," *Journal of Economic Perspectives* (Spring 2004) 18:2 pp. 107–126.

8 Laitner, John. *Energy Impact of Emerging Technologies* (Washington, DC: Environmental Protection Agency, 2004).

9 Halal, *The New Capitalism: Business and Society in the Information Age* (NY: Wiley, 1986).

10 See "The GW Forecast of Emerging Technologies," *Op. Cit.* and "Emerging Technologies," *The Futurist* (Nov–Dec 1997); reprinted in O'Meara *et al.* (eds) *Globalization and the Challenges of a New Century* (Indiana University Press, 2000).

11 See my special issue of *On-the-Horizon*, "Institutional Change" (2005).

12 Halal, "The Coming Conflict Between Science and Spirit," *Thinking Creatively in Turbulent Times* (2004).

13 Halal, "The Life Cycle of Evolution," *Op. Cit.*

14 Among the most prominent of these modern day prophets is Willis Harman, author of *Global Mind Change* (Indianapolis, IN: Knowledge Systems, Inc., 1988).

Part I

Forecasts of the Technology Revolution

CHAPTER 2

Transition to a Sustainable World: Converting the Energy and Environment Mess into an Opportunity

The energy and environment problem is one of those nagging old dilemmas that continue to haunt us because we rarely see the enormous opportunities it conceals. Lee Scott, CEO of Wal-Mart, admits he started his company's conversion to sustainability as a "defensive strategy" but it "turned out to be precisely the opposite." Now the program has Wal-Mart employees feeling good about their company, and saving customers money while protecting the environment.[1]

THE ECOLOGICAL CHALLENGE

As the next chapter will show, industrialization is likely to spread over the globe during the next 20 to 30 years, but it will require a rather complete overhaul of our economic systems, technologies, politics, and cultures. The stark facts make this almost inevitable. A little more than one billion people now live in industrialized societies, but the global population will reach eight billion about 2030. Most will wish to live at modern standards because they want to enjoy the modern comforts they know we have, and globalization will give them the chance to do so. The result is that industrialization will produce at least a three-fold gain in material consumption, energy use, and environmental load over the next two or three decades, and possibly as much as a five-fold leap. China alone will double these problems, and India will do the same.

The global environment is in precarious ecological balance even now, so how could we possibly tolerate a three- to five-fold increase of industrialization? With oil supplies tightening and fears of climate change, the energy and environment problem has to be resolved over the next 20 years or so if civilization is to survive. Enter the words "peak oil" on Google, and you will find more than one million warnings of the calamity that may await if everything we depend on – from travel to plastics – is threatened by spiking oil prices.

Russell Ackoff defined a constellation of interrelated problems like this a "mess" and Al Gore called it "An Inconvenient Truth."[2] There is no avoiding the uncomfortable reality that we are living beyond our means and the problem is increasing. The energy and environment mess is so fundamental to everything else that it deserves top priority. That's why it is the first field in the TechCast system. Because a sound environment is the foundation of life, without it all else is moot.

With some good judgment and a little luck, we are likely to muddle through as threats of disruption spur adaptive change. Global oil production seems to be peaking this decade, although there is controversy, but the peak is likely to be a "rounded top" that may last 5 to 15 years during which the world transitions to a sustainable economy. We are even now seeing hybrid cars, wind turbines, biomass, more nuclear plants, solar power, and conservation replace carbon fuels. The ominous warnings of imminent global disaster to be caused by the Y2K problem fizzled for the same reason – the warnings themselves were so dire that they forced corrective change.

Below is summary of what our experts think. As noted in the Introduction, references to facts, quotes, and other material are available on the website www.TechCast.org.

Summary of Energy and Environment Forecasts

Alternative Energy Wind, hydro, solar, biomass, and other alternative sources are likely to increase from the present 17% level to 30% of all energy by about 2020.

Desalination Rapidly declining costs will provide good water treatment at modest prices, solving one of the most critical resource issues threatening the globe.

Distributed Power The electric power system is decentralizing into a robust network of suppliers, businesses, and homes feeding the grid with excess local energy.

Genetically Modified Organisms Despite fierce opposition, designed strains of plants and animals are improving farm yields and offer enormous other possibilities.

Global Warming The evidence is mounting that it is necessary to limit greenhouse gases, so governments are starting to take serious actions to forestall climate change.

Green Business Corporations are starting to embrace sustainable practices because they are now seen as a competitive advantage, leading to an economic boom.

Organic Farming With Wal-Mart embracing organic produce, the elitist stigma of organic is yielding to good marketing and a growing concern for healthy foods.
Precision Farming Farmers are moving to high-tech because it optimizes farming practices to lower costs, raise yields, and minimize ecological damage.

Alternative Energy

The need for alternatives to oil is severe because global oil production may be peaking at a time when developing nations are using more energy and concern over the environment is growing, signaling the end of a long era dominated by carbon fuels. Oil, gas, and coal presently supply 83% of all energy use, with the rest obtained from hydro – 6%, biomass – 3%, nuclear – 6%, and wind and solar – 2%. Therefore *renewable* energy (excluding nuclear) totals 11% and *alternative* energy is 17%. Alternative sources are growing 30% per year, backed by wide public support and corporate investment. As oil prices continue to rise and the cost of alternatives falls, our forecasts suggest that carbon fuels may no longer be the main energy source in two to three decades.

Carbon fuels may remain plentiful because new oil fields and extraction methods are constantly being discovered. The U.S. reserves consist of mainly of coal, and the gasification process may soon produce clean liquid coal while the CO_2 is sequestered in oil wells where it could help recover more oil. The oil sands of Canada and the Western U.S. are enormous and could meet demand for decades. The U.S. Geological Survey and the Department of Energy think world oil production will last for 40 years.

But the same method that correctly forecast U.S. oil production to peak in the 1970s now indicates global supplies will peak this decade.[3] And economic trends favor alternatives because rising demand raises oil prices,[4] while the cost of renewables is dropping. In 2006, the cost of wind and hydro matched carbon fuels, while biomass, solar, and nuclear were almost competitive with oil. Additionally, the environmental costs of carbon fuels are climbing. Surveys show that 80% of people around the world want to protect the environment and 75% support renewable energy.

With this shifting economic situation, business is embracing alternatives because they now offer huge growth opportunities. Clean energy companies attracted 1% of venture capital before 1999 but are now getting 5% to 8% of all investment. Entrepreneurs say "Greentech" could be "the largest business opportunity of the 21st Century," and "We're seeing the same ramp up we saw with the web." Here's a summary of developments:

Alternative Energy Sources

Wind Experts estimate that wind power could supply 12% of the world's energy by 2020. England is building offshore wind farms housing 6,000 turbines, and Germany is building 16,000 turbines. Germany and Denmark expect to derive half of their energy from wind. Wind turbines in North Dakota, Kansas, Texas, and other U.S. states alone could provide a large portion of America's energy needs.

Solar Entrepreneurs around the world think solar cells will produce electricity at prices that compete with oil in a decade or less, causing the market to explode.[5] Photovoltaic cells cost $3/watt to $4/watt to build, compared with $0.40/watt for oil and gas. A TechCast expert, Yair Sharan, estimates that total solar energy costs, including usage, to be about 20 cents to 40 cents/kWh versus 4 cents to 12 cents/kWh for oil. But nanotech can provide plastic solar cells at $0.20/watt and increase efficiency. Nanosolar Company is mass producing solar cells at far less cost by simply printing them, and expects to increase the global supply 20-fold. The world's largest solar power plant, located in the Mojave desert, is 30% efficient. The CEO says that "11 square miles could produce as much energy as Hoover Dam."[6] The consensus is that costs will become competitive with other energy sources about 2012 to 2015, and some experts estimate solar and wind power will reach 10% of U.S. energy by 2013.

Fission Nuclear fission produces no pollution, has a sound safety record, and compares with oil on cost. About 70% of Americans now favor nuclear energy, and 88% of those living near nuclear plants are confident of their safety. New designs like the pebble bed reactor are immune to meltdown, and waste can be stored safely if coated in glass or socked away in stable mountains for 200,000 years. A "fast breeder" design recycles spent fuel to reduce radioactive waste from 95% to 1% of the fuel used, increasing efficiency and reducing the disposal problem. The Bush Administration plans to use these technologies to build a new generation of reactors that reprocesses spent fuel, thereby eliminating nearly all waste, extracting 100 times as much energy, and reducing greenhouse gases. Worldwide, the number of nuclear power plants is expected to grow from 435 today to 600 by 2010. In anticipation, the price of uranium rose from $11 per pound in 2003 to $136 per pound in 2007.

Biomass With gas prices in the U.S. reaching $3 per gallon, biomass is poised to takeoff as it becomes economically competitive. Capacity for producing ethanol from corn has doubled in the last three years and will double again soon, despite its low efficiency and the tightening of corn markets. Cellulosic ethanol is being made from woody plants and biomass waste (sawdust, weeds, etc.) at lower cost and better efficiency. The Environmental Protection Agency (EPA) estimates that biomass energy could replace 40 billion of the 110 billion gallons of oil the U.S. uses each year.

Conservation The U.S. economy operates at such low energy efficiency that conservation is the most cost-effective approach to supplying our future needs. The U.S. saved 40% in energy costs over the past 20 years using simple forms of conservation. Amory Lovins thinks conservation can reduce energy costs by 75% with a 100% annual return on investment.

Innovation Light Emitting Diodes (LEDs) could save 90% of the energy wasted by incandescent lights. Scores of new fuel cell technologies are being developed to create H_2 directly from biomass. Photosynthesis is becoming understood, offering the prospect of converting sunlight into energy as plants do, at 100% efficiency. Tidal energy is being harnessed in Manhattan, France, and Nova Scotia. Geothermal energy is producing the first hydrogen economy in Iceland. Cold fusion is being re-examined because of new supporting evidence. Researchers at the University of California are converting the biggest problem in global warming – CO_2 – into oxygen and carbon monoxide, the primary feedstock for plastics and other products. Wind turbines are being developed that ride 10 km up in the jet stream to capture 100 times as much energy, which is transmitted to Earth on supporting cables. The U.S. military and India are studying the use of solar satellites for producing energy.

The trend is unmistakable. California Edison increased its use of renewables from 1% in 1985 to almost 30% today, and California now requires 20% of its energy to be renewable by 2017. Maine now derives 30% of its energy from hydroelectric plants and other renewables. Hawaii, and other states aim to produce 20% of all energy from renewables by 2020. The U.S. Department of Energy (DoE) thinks renewables will reach 28% by 2030, and the E.U. expects renewables to reach 22% of energy use by 2010.

After reviewing the evidence, TechCast experts estimate 30% of global energy will be derived from alternative sources about 2023

+/– 5 years, beginning the transition from carbon fuels. When the transition to alternative energy is well underway in 30 years or so, we estimate hydro will continue to provide the same 6% of global use, biomass will grow from 3% to roughly 10%, nuclear will grow from 6% to about 9%, wind will increase from 1% to about 10%, solar will grow from less than 1% to as much as 20%, and conservation will produce about 10%. The remaining 35% or so is likely to be come from oil and gas. It could not happen soon enough.

Desalination

The need for clean water is severe in most parts of the world and growing. Water use worldwide has grown six-fold since 1900, yet half of the world suffers from unsafe water systems. Five million people die each year because of unsanitary water. *Fortune* said, "Water promises to be in the 21st century what oil was in the 20th century."

Desalinization was costly, but technical advances are solving the problem. For instance, a "capacitive deionization" technique produces clean water at half the cost of the conventional reverse osmosis technique. Nanotech is being used to create filters that block bacteria and viruses. "Rapid spray evaporation" technology cleans water cheaply and produces no brine byproducts. Overall, costs have dropped from $20 per gallon in 1950, to $6 per gallon in 1960, and are now approaching 1 cent per gallon. Ovation Products claims it can distill water contaminated with anything into pure drinking water for 1 cent per gallon.

Although most water purification systems were managed by government monopolies, private firms are moving in to encourage innovation. The French firms Vivendi and Suez are the largest water treatment companies in the world and are growing rapidly. California is building 13 plants that could supply 20% of the state's water, Florida is building the largest plant in the Western hemisphere, and Texas is planning nine plants. Israel has five large projects. We expect desalination to enter mainstream use about 2020.

Distributed Power

Prominent failures of today's centralized power systems have heightened interest in distributed grids that are more reliable. The 2003 blackout in the Northeast U.S. was the fourth catastrophic failure in a decade. It was followed in September by another blackout in Scandinavia. Distributed grids are self-managed networks relying primarily on local energy sources that can reduce transmission costs, the risk and severity of failures, and vulnerability to attack.

However, local power sources are not well developed as yet. Only 500 homes in the U.S. are defined as "zero energy" because they derive 70% or more of their power locally. But one million U.S. homes use some degree of local energy, and builders are installing solar panels in new homes, usually costing $4,000 to produce 1 kW. Forty states let people sell power to the grid. Germany and Japan are well ahead of the U.S. in building zero-energy homes.

Some analysts expect distributed power to provide 20% of all U.S. electric energy by 2010, while John Benner of the National Renewable Energy Lab thinks that local power sources will supply 25% of new capacity by 2020. TechCast experts estimate that the amount of power derived from distributed grids is likely to increase from the present level of 7% to 30% by about 2021 +/– 5 years.

Genetically Modified Organisms (GMO)

GMO is the convergence of biotech and agriculture to design strains of plants and animals that grow better and have more useful properties. More than half of all U.S. corn, soybeans, and cotton farming uses genetically altered seed. GMO is used on 20% of U.S. farmland, but only 1% of global farmland. The public fears manipulation of the codes of life, so testing is underway to ensure the safety of GMO crops, and accurate labeling is needed before this sensitive technology is accepted widely. Despite these concerns, the benefits are so great that the obstacles are likely to be overcome in time. A Monsanto executive described the possibility of "using plants as factories to produce almost anything."

Greenpeace and a consortium of 650 farming groups filed suit to force the U.S. to ban genetically altered food, and a majority of Britons and Europeans reject the concept. Even in the U.S. where the public seems complacent, one survey found that 38% of Americans would use genetically altered food, while 56% would not. "The industry is in big trouble," said an agricultural consultant. "They overstated the science." Other experts called negative public attitudes "a triumph of fear over reason."

These are serious concerns, but GMOs offer endless possibilities. Dow Chemical, Monsanto, DuPont, and other companies have invented a dozen or so strains of plants that grow faster, resist disease, fertilize themselves, and produce valuable nutrients. Pharmaceutical companies are designing food called "neutraceuticals," "biopharming," or "therapeutic foods," able to treat medical illnesses. For instance, tomatoes can produce anticancer ingredients, eggs can produce cholesterol-lowering substances, and other plants yield the omega acids that make eating fish healthy. A genetically engineered bacteria now mass produces the drug "artemisinin" to cure malaria at 25 cents per treatment, and rice containing proteins can combat diarrhea in children of poor nations. Even though the Amish are suspicious of technology, the benefits convinced them to use GMO. "We'll see massive changes in managing disease and nutrition," said a DuPont VP.

The need is huge. World population growth on limited farmland requires continual increases in crop yields to feed a projected increase of three billion people, even while hunger is a serious problem in parts of the world. An African official said, "We do not want to be denied this technology because of a misguided notion that we don't understand the dangers."[7] Traditional breeding has long been used to alter genetic strains of plants and animals and to produce drugs (insulin) and enzymes (yeast), so GMO are not really new. What is new is the larger scale permitted by bioengineering.

In this struggle between hope and fear, progress is slow but steady. Britain approved its first use of genetic food, considered a symbolic event for the E.U. Within five years, ten million farmers in 25 nations are expected to plant 100 million hectares of genetically modified food. Europe passed laws requiring labels, and New Zealand requires testing, public approval, and labeling Given the political uncertainties, however, TechCast thinks it could take 10 to 20 years to reach the 30% adoption level.

Global Warming

Popular opinion, most scientists, and even corporate CEOs have accepted the dangers of global warming. The International Panel on Climate Change, composed of hundreds of scientists from 113 nations working in 18 separate groups, recently announced. "There is an overwhelming probability (> 90%) that humans are warming the planet at a dangerous rate."[8]

The fact remains, however, that the rise in global temperature over the past century is within the range of previous cycles, encouraging a small but adamant minority to insist that the warming trend is natural. Despite the correlation of global temperature with rising CO_2 levels, despite several unusually warm years in the past decade, despite extraordinary heat spell in Europe that killed thousands, despite the polar ice caps melting, some think this is just a blip in the Earth's normal rhythms.

If the current warming is part of a natural cycle, cutting back on oil is not likely to reduce global warming, it could incur huge costs, and waste needed energy. Even if the Earth continued to warm, some scientists contend this would be beneficial because crops grow better in a high CO_2 climate. Regions near the equator would suffer, but others closer to the poles (Canada, etc.) would benefit. It's easy dismissing skeptics as cranks, but some have serious doubts. Freeman Dyson, a great scientist, thinks the "The fuss ... is grossly exaggerated."[9]

The Kyoto Protocol is inadequate, so there is no working agreement for lowering greenhouse gases. An E.U. study forecasts a doubling of greenhouse gases by 2030 as developing nations burn large amounts of coal. China, India, and the U.S. are building 850 coal-fired plants, adding five times as much CO_2 as the Protocol intended to reduce.

The problem could escalate to catastrophic proportions before action is taken. Scientific models indicate that changes in CO_2 levels require 50 years or more to affect global temperatures because the Earth is such an enormous heat sink. Present concerns are the result of a 1° F rise in global temperature, but the increase could be 10° or more by 2100. A study in *Nature* estimates that global warming could destroy a quarter of all animal species, the greatest extinction since the dinosaurs. To make things worse, the effect of greenhouse gases has been limited by sun-blocking smoke particles, so temperatures could rise faster with pollution control.

Perhaps the truth lies somewhere between both arguments. This could be a natural cycle exacerbated by carbon fuels, raising the question: *To what extent* is global warming "man-made" and therefore capable of remedies? TechCast surveys show knowledgeable people think humans cause about 70% of global warming, and a Fox News poll found 77% of Americans think global warming is a serious problem.

Thankfully there are signs of progress. Britain has trimmed emissions by 15% since 1990 while its economy grew 36%, demonstrating the feasibility of pollution controls; the nation expects to reduce emissions another 60% by 2050. California now requires that greenhouse gases be reduced to their 1990 levels by 2020. Major companies (GE, Ford, IBM, DuPont, etc.) have each reduced emissions by 60% since 1990 while saving billions in the process. (See Green Business.) China is rapidly becoming the largest economy in the world, but with environmental problems so severe the nation is actively planning a "circular economy" involving "cradle to grave" recycling. TechCast estimates the problem of global warming will be addressed seriously by most nations in the next decade, most likely around 2014 +/– 5 years.

Green Business

Although sound environmental management remains rare, many companies are making serious efforts because green practices are economically advantageous, they convey a favorable public image, and technical advances are making them cheaper and more practical. In addition to Wal-Mart noted above, the list of green companies reads like a *Who's Who* of the *Fortune* 500:[10]

Exemplars of Green Business

DuPont has reduced greenhouse gases 50% because "we found it pays to be ...in a better competitive position."

Dow Corning made $1 billion using green methods that save resources, decrease taxes, and create valuable byproducts.

Johnson & Johnson redesigned packages to save 3,000 tons of packaging and 2,000 tons of paper at a savings of $3 million, while protecting 330 acres of trees.

Carrier now sells a "comfort level," using insulation to save energy, rather than simply installing air conditioners, reducing costs and increasing profits.

Ford rebuilt its large River Rouge factory in Detroit by installing skylights and a "living roof" of sod that reduces toxins, boosts productivity, and saves money.

Genzyme Corporation designed its new building with a sod roof, skylights, and blinds, and it uses waste steam for heating.

The New York Jets stadium was designed to use solar cells and wind turbines to reduce energy, selling excess energy to the city's electrical grid.

UPS and FedEx are equipping vans with hybrid and fuel cell engines to reduce fuel costs and pollutants.

Dow Chemical expects 25% of revenues to come from renewable resources by 2010.

And governments are urging them on. The E.U. requires manufacturers to recycle their products. The 178-nation Kyoto Protocol encourages trading in pollution credits, which is now turning into the biggest commodities market in the world. Maryland requires chicken growers to manage chicken manure, so Perdue Company built a plant that converts it into fertilizer under the brand name "Cockadoodle Doo."

Considering all the evidence, TechCast forecasts that 30% of corporations will practice progressive environmental management by 2011 +/– 4 years.

Organic Farming

Despite millennia of natural cultivation, organic farming is one of those promising ideas tainted by the belief that it offers unrealistic claims. Experiments have demonstrated the advantages, but critics say that artificial chemicals are essential to feed a growing world. One analyst said "Organic farmers are wonderful people, but they oversell." Farmland is being damaged by the continued use of chemicals, however, and concerns about public safety support organic methods, so the practice is growing quickly. Wal-Mart recently decided to sell organic produce at just 10% above normal cost, and their marketing clout could make organic foods common. One authority said "Wal-Mart has legitimized the market."

The basic problem is that chemical fertilizers and pesticides destroy micro-organisms in soil and develop resistance in pests, creating a vicious cycle in which yields drop with time and costs increase as more chemicals are needed. Studies find that organic farming uses less resources, maintains high crop yields, improves the soil, produces higher profits, reduces drought and erosion, and protects the environment. Studies also show that organic foods produced a "dramatic and immediate" drop of pesticide levels in the bodies of children. Research shows that "factory" raised chicken has more fat and cholesterol and less vitamin A and omega acid than "free range" chicken, and grain-fed beef show similar declines compared to pasture-raised animals. Vegetables also lose nutrients when grown with chemicals. Surveys show 90% of the public favors organic produce.

Organic farming represents a difficult break from modern methods to a labor intensive approach that must be adapted to the unique needs of each parcel of land. But it suits the family farm well. An agriculture official said "Alternative farming could transform the industry and make it profitable to save the family farm." Organic comprised 2.5% of farming in the U.S. and 3% in Europe as of 2005, but it is growing by 30% per year, the fastest growth sector in the food business. Europe is expected to produce 10% to 20% of its food organically by 2010. TechCast studies estimate 30% of farmland is likely to be cultivated using organic methods about 2022 +/– 6 years.

Precision Farming

Precision farming (PF) involves the computerized control of irrigation, seed distribution, fertilizer, and pesticides to suit variations in land identified using Global Positioning System (GPS) and geographic information systems. Twenty percent of farmers in the U.S. are adopting these practices because PF reduces the amount of costly chemicals needed, raises yields, and protects the environment. A farm equipment manufacturer said, "In 10 years, every farmer will use this technology."

PF is being developed mainly by large corporate farms that have the investment capital and management skills. Farm equipment manufacturers like Case and Deere sell PF systems on tractors for about $6,000, and they provide farmers data warehouses of critical information to guide their operations. The practice has grown as farmers become impressed with the higher yields and lower costs. A review of 108 studies found 63% of applications were more profitable than farms using traditional methods. TechCast estimates PF to reach mainstream use about 2010 to 2015.

THE TRANSITION TIMETABLE

To better understand the implications, let's review these technologies in strategic terms using the bubble chart (Figure 2.1).

Focusing on the upper left corner highlights where high confidence and large economic markets suggest significant breakthroughs are imminent. We can expect business to create an economic boom as

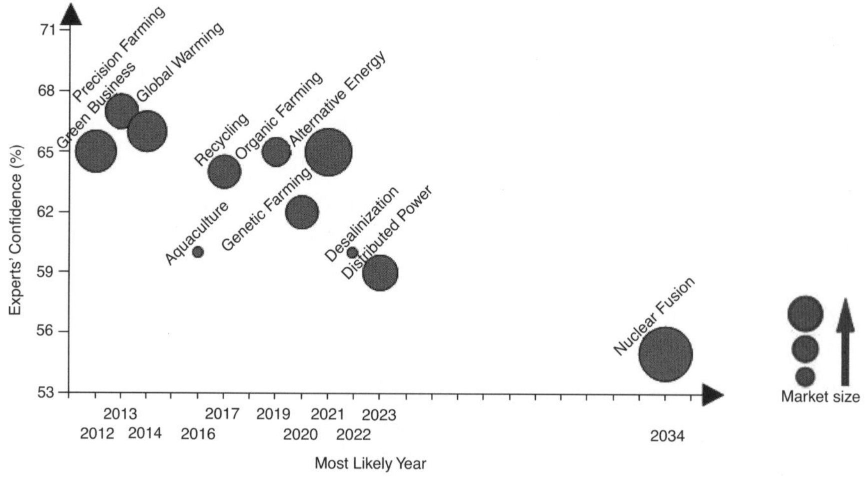

Figure 2.1 Bubble Chart for Energy & Environment

green practices move into the mainstream over the next five years or so. Moving toward the right in time, the decade of 2010 should prove critical to address global warming, which would also help to transition toward alternative energy by about 2020 and encourage a distributed power grid.

This suggests the transition to sustainability is beginning, although with the normal level of doubt and cost that accompany such historic change. We don't know the details, but we do have a rough timetable of how and when this transition will occur. The evidence shows that modern economies are adapting to these new realities with a wave of innovative new energy sources, many tucked into the interstices of society – hybrid cars, solar panels on roofs, windmills on a farm, ethanol plants in Iowa, and nuclear power plants where they are wanted.

Economically, green practices usually prove to be rewarding investments. What is so striking is that almost all soundly conceived projects show unusual gains through lower costs, resources saved, public goodwill, preserving environmental assets, and making money. That's why a huge new industry devoted to environmental practices is taking off to become one of the most crucial economic sectors. The world market for pollution control was $500 billion in 2000 and is expected to reach $10 trillion in 2020,[11] larger than automobiles, healthcare, defense, and other big industries.

Not only is the environmental mess a great opportunity in disguise, it also encourages the type of collaboration so badly needed. Starting an ordinary business is fraught with conflicting interests, legal issues, competition, and other ordeals. But there is something about the environment that is unifying. Once the problem is addressed squarely, everyone can usually gain through creative solutions. The classic example is Interface Corporation, a $1.1 billion company that provides "carpet service" rather than selling carpet itself. They learned to recycle carpeting, and found it lasts four times longer and uses 40% less fabric, while reducing the amount of replaced carpeting by 80%. Overall, this produced a 35-fold reduction in materials. Ray Anderson, the CEO, said, "Sustainability doesn't cost, it pays... Our costs are down. Our products are the best they've ever been. Our people are motivated by a shared higher purpose. And the goodwill in the marketplace ... has been astonishing."[12]

In addition, there is a higher calling associated with protecting the Earth. After the value of Wal-Mart's environmental program became clear, the CEO said to his employees, "Doesn't it feel good to have this kind of commitment made by the company that you are part of? Don't you feel proud?"

The present situation offers a great opportunity to cut the Gordian Knot, converting a potential crisis into sustainable, unifying growth. The prospects for this infant industry are so great they justify a "Green Manhattan Project." The U.S. government should invite major corporations and other governments to form collaborative projects for improving environmental management, alternative energy, and other sustainable technologies. These same groups should agree on a system of carbon taxes to internalize the costs of producing greenhouse gases and allow the market to solve environmental problems more efficiently. We also need to encourage innovative solutions, like sequestering CO_2, planting trees, and using industrial ecology. By acting aggressively, the U.S. could restore its tarnished reputation as a world leader.

Our forecasts suggest the world is likely to realize the benefits of ecologically safe living during the next 10 to 20 years. A rising interest in protecting the environment is beginning to integrate industries, energy systems, farming, homes and offices into a living tapestry that sustains life. Amory Lovins and Paul Hawken call it a "Natural Capitalism" in which the environment is recognized as a valuable asset that produces \$33 trillion of various economic benefits annually.[13] The challenges are enormous, but they are being resolved and the timetable of the path ahead is fairly clear. Now we need to improve the technology, implement it widely, and find the political will.

Notes

1 Marc Fisher, "The Green Machine," *Fortune* (July 31, 2006).
2 Ackoff, *The Democratic Corporation* (NY: Oxford University Press, 1994).
3 Deffeyes, *Hubbert's Peak: The Impending World Oil Shortage*, 2003. Also see, Matthew Simmons, *Twilight in the Desert: The Coming Saudi Oil Shock and the World Economy* (Hoboken, NJ: Wiley, 2005).
4 Some forecasters think oil prices will exceed \$100/barrel soon and stay there. See "\$100 Oil Price May be Months Away," *CIBC World Markets* (July 23, 2007).
5 See "Investors Look to Future in Renewable Energy," *Washington Post* (August 26, 2007).
6 Also see Travis Bradford, *Solar Revolution: The Economic Foundation of the Global Energy Industry* (Cambridge, Mass: MIT Press, 2006).
7 Hassan Adamu, "We'll Feed Our People As We See Fit," *Washington Post* (Dec 20, 2005).
8 *Scientific American* (August, 2007).
9 Dyson, "Heretical Thoughts About Science and Society," *Edge* (August 10, 2007).
10 Also see *Green to Gold: How Smart Companies Use Environmental Strategy to Innovate, Create Value, and Build Competitive Advantage* (New Haven, CT: Yale University Press, 2006).
11 Al Gore, *Earth in the Balance* (NY: Houghton-Mifflin, 2000).
12 Cornelia Dean, "Executive on a Mission: Saving the Planet," *New York Times* (May 23, 2007).
13 Paul Hawken, Amory Lovins, and L. Hunter Lovins, *Natural Capitalism: Creating the Next Industrial Revolution* (Snowmass, CO: Rocky Mountain Institute, 1999).

Chapter 3

Globalization Goes High-Tech: A Worrisome World of Abundance

I was visiting the carefully preserved medieval ruins underlying Turku, Finland, recently, and the remains of these small stone and wood homes forcibly reminded me of the harsh way people lived just 200 hundred years ago. Before the Industrial Revolution gave us the comfortable lives we enjoy today, the average European lived in a large hut of some type, often with dirt floors and no windows. Indoor plumbing? Electricity? Medical care? Refrigeration? All yet to be invented. Economics was originally defined at about that time as managing *limited* resources that *decrease* when shared in a world of *scarcity*. Little wonder it was called "the dismal science."

In contrast, today's economic system based on knowledge is changing the way the world works. Knowledge is *infinite*, it can *increase* when shared, and creates *abundance*. When the late Ray Smith was CEO of Bell Atlantic, he called it the "loaves and fishes" principle: "Unlike raw materials, knowledge can't be used up. The more you dispense, the more you generate."[1]

For example, consider the abundance produced by the Internet, that quintessential structure of the Information Age. Google, Yahoo, and Microsoft give away email accounts with huge storage capacity. Anyone wanting to find their way to almost anywhere can get free directions and maps on MapQuest. Wikipedia uses volunteers to offer constantly updated reference material. It is among the most popular sites on the web, with more readers than the *New York Times*, the *Washington Post*, and the *Wall Street Journal* combined.

The growing power of knowledge explains why high technology is poised to spread around the world. The age of belching factories and capitalist titans is yielding to more complex business systems powered primarily by knowledge in its various forms. As this chapter will show, some of the most crucial innovations today involve sophisticated materials and machines able to do almost anything; far more intelligent equipment, robots, software, and people; all working online to unite the globe in a web of buyers, manufacturers, and suppliers.

Physical manufacturing systems and capital remain essential, of course, and globalization is growing even now as the newly connected world becomes "Flat," in the words of Tom Friedman.[2] But far more sophisticated forms of economic growth are spreading rapidly because the high-technologies emerging now are basically knowledge that can be created, passed on, and applied widely. The following innovations promise to create new industries and services that will industrialize the globe.

TECHNOLOGIES OF ABUNDANCE

The seven technologies highlighted below outline how this revolutionary transition is likely to transform economies over the next 10 to 15 years. As before, please remember that references to the many facts, quotes, and other data noted here are available at www.TechCast.org.

Summary of High-Tech Manufacturing Forecasts

Designed Materials Understanding the deep structure of matter is making it possible to create almost any conceivable type of material product.

Mass Customization Intelligent IT systems are automating production and distribution of almost perfect goods, customized to order, cheaply.

Micromachines Tiny machines can control air bags, light projectors, examine the body's interior, and go where others cannot.

Modular Homes Homes themselves are being reconceptualized as modular systems built in days to any conceivable arrangement.

Nanotechnology On an even smaller scale, the nano world offers high-performance materials, powerful computers, medical treatments, and other benefits yet to be realized.

Smart Robots All these advances are combining to produce smart robots working in factories, offices, and homes as assistants, servants, and caregivers.

Smart Sensors Tiny devices can monitor almost anything, wiring organizations and the entire globe into a web of collective intelligence.

Designed Materials

Materials research is increasingly able to design products with great strength, adaptive behavior, and other unusual qualities. A scientist at the U.S. National Science Foundation thinks it is now possible to create "anything you can think of."

For instance, carbon fiber-laced composites can be made stronger than steel, biodegradable, able to conduct light and electricity, to withstand high temperatures, resist corrosion, and cushion impact better than honeycomb. Amory Lovins contends that composite car bodies would cut oil use by 30% and recover costs in three years. "Electronic paper" composed of tiny half-black, half-white balls can display moving images. Intelligent polymers that respond to temperature are being used as surgical sutures that morph into perfect knots when exposed to the patient's body heat. Spider's silk has been synthesized to create fibers that are three times tougher than other synthetic fabrics. A German research lab has developed a transparent form of aluminum, "alumina," that is extremely light but three times tougher than hardened steel.

Nanotech and biotech serve as foundations for materials research, and they are improving rapidly, so a wider range of applications should become common in aircraft, automobiles, and other industries. The forecast data suggest some uncertainty, but our most likely estimate is that designed materials will enter the mainstream about 2015.

Mass Customization

Progressive companies are taking customer orders online and using automated production methods to deliver high quality, personalized goods at lower prices. "Mass customization will proceed with gravitational force," said Alvin Toffler. "People desire greater individuality as they become more affluent, and technology allows it." Noted experts Joseph Pine and Stan Davis wrote, "Mass customization will be as important in the 21st Century as mass production was in the 20th Century."[3]

As of 2007, online selling in the U.S. represented only about 6% of retail sales, but customization is likely to grow because it satisfies the unique needs of a diverse population. Companies now offer customized PCs, jeans, shoes, cars, bicycles, computers, and other goods in various colors, sizes, materials, and other options to suit different tastes. Online retail also is being used to offer a diversity of financial services, interactive TV, music, hotel services and other intangibles. Callabray produces 3D computer models of each client's body to customize clothing, and Saturn invites people to configure their new car online. Managers say "Our goal is simultaneous manufacturing; to make products as the customer talks," and "All marketing today is going to customization. It's the secret to customer retention."

The underlying technologies that support customization – e-tailing, broadband, computerized manufacturing – are all expected to reach critical mass about 2010. Because mass customization can be used on everything from toothpaste to insurance, it should have a dramatic impact on economies, spurring growth, paring costs, increasing convenience, and reducing environmental impacts. A variety of forecasts point toward 2010 to 2016 as the take-off period when mass customization is likely to enter the mainstream of modern nations.

MicroMachines

Tiny machines, or Micro-Electro-Mechanical Systems (MEMS), can be made using the same photolithography process used to create silicon chips. They are being used for air bag actuators, digital light processors, optical switches, inkjet print heads, implantable medical dispensers, and pressure sensors. "We're on the verge of a second silicon revolution," said a scientist at Sandia Labs.

Consider some of these marvelous little devices: Researchers have developed a microturbine jet engine the size of a dime that runs at 1 million RPM and delivers 20 to 50 watts to power cell phones and laptops. Texas Instruments has developed a digital light processor consisting of 20 million small mirrors that produce brighter light for movie projectors. Air bags are controlled by MEMS consisting of a tiny sensor and valve system, about the size of a pencil eraser. A microscopic probe is available that can be guided wirelessly through the esophagus, stomach, and colon to detect cancers. IBM has developed a data storage system (the "millipede project") that uses tiny fingers to punch holes in a disc to achieve data densities of 1,000 gigabits/square inch. Radio-equipped robots smaller than a grain of sand can move through the air or water to monitor and control anything. It could soon be possible to inject MEMS into the circulatory system to battle illness by destroying dangerous organisms.

We estimate such devices should become common about 2020.

Modular Homes

Associated with cheap design for decades, prefabricated or modular homes are becoming a stylish, cost-effective alternative to onsite construction. "Prefab was thought of as bad design," said architect Jill Herbers. "Now it's the polar opposite." Leading architects in Europe, the U.S., and Asia are designing modular systems that can be ordered in any combination to erect a unique home for each family that is attractive, functional, and inexpensive.

Typical modular homes cost half or less the price of traditional homes and can be constructed in days, although buyers often provide the foundation, wiring, plumbing, and heating. For instance, a home designed by architect Rocio Romero is a kit delivered on a truck for $30,000, while Dwell Homes average about $50,000 for a two-level family home. Architect Kent Larson plans to allow homes to be designed on the Internet by choosing modular components, with standardized systems for wiring and plumbing. The Swedish furniture store IKEA is selling entire homes for about $100,000 that can be assembled much like IKEA furniture. A robotic machine can guide nozzles that lay concrete in layers to form any shape of construction in 24 hours at 20% of current costs; the inventors were inspired by the use of inkjet printers to create 3D prototypes out of plastics and other materials. They say, "Anything you can dream, we can build."

Modular construction will have to overcome the traditional customs and practices associated with stick-by-stick construction. With globalization creating huge demands for affordable housing in poor regions of the globe, however, it's easy to imagine a ground swell of modular design changing homes forever. Modular systems only account for about 15% of construction today, but TechCast estimates the practice will reach 30% use about 2015.

Nanotechnology

The "nanosphere" consists of objects measured in one-billionths of a meter, and it is now undergoing a revolution as research increasingly yields control over this tiny world. Things behave differently at the nano level, with electricity flowing 100 times more easily and materials changing properties. The possibility of creating more powerful computers, medical treatments, and almost any type of extraordinary item from scratch has the entire world excited, with billions of dollars being devoted to research that produces breakthroughs daily. Some 1,200 nanotech ventures have been started around the globe, and 3,000 nanotech patents have been filed. The President of NanoBusiness Alliance claims "Nanotech research is breaking out all over the planet."[4]

Samsung, Motorola, and other companies are now selling supersharp flat-screen displays using nanotech to project high-fidelity images at a fraction of the power of Liquid Crystal Display (LCD) screens. National Aeronautics and Space Administration (NASA) gave Rice University a $1 million grant to grow nanotubes into electrical wires that will carry power with one-tenth as much resistance as copper. Ultracapacitors made with nanotechnology can outperform batteries; rather than store energy chemically, it is stored as a charge between plates constructed of nanotubes. Because they are so small, nanotubes increase the surface area enormously, holding more energy, recharging in seconds, and lasting indefinitely.

Various advances could produce nanotech computers that outperform silicon chips. Nantero and ASM Lithography even now are producing nanotube memory chips that store terabits of data per square centimeter, about a million times the current data densities. Processors are harder to create, but IBM has used carbon nanotubes to form transistors 100 times smaller than those on silicon chips. Magnetic nanoparticles are being used to create instant-on computers that retain data without power.

It may seem impossible to organize something as small as nanotubes into such intricate patterns, but they form spontaneously in a self-correcting process that is predictable and error-free. The power of these developments is so great that they have been called "the biggest breakthrough since the transistor."

Another area of huge potential is nano-medicine. The U.S. National Cancer Institute has launched a $144 million program to "radically change the way we diagnose, treat, and prevent cancer." As we will see in Chapter 5, the focus is on creating nano-sized devices that can seek out dangerous cells and destroy them, somewhat like scavengers. The U.S. National Science Foundation estimates that half of all medical treatments and drugs could be affected by nanotech.

One of the most striking applications is the space elevator. A few decades ago, a Russian scientist determined that it was feasible in principle to place a geosynchronous satellite in a 62,000 mile-high stationary orbit, drop a cable down to Earth, and ride up and down to space on an elevator. The only problem was, where to find a cable that could support 62,000 miles of its own weight?

The unusual strength of carbon nanotubes – 100 times the tensile strength of steel at one-sixth the weight – has now made that vision possible. Scientists at Los Alamos Laboratories are designing a space elevator that will orbit above a fixed spot on the equator, connected to earth with a thin film of nanotubes 1 meter wide. The LiftPort Group, a private venture, plans to build a functioning space elevator. They are now experimenting with a mile long cable tethered from balloons. "It was rock solid," said the president of the company.

TechCast thinks nanotechnology is likely to reach mainstream use about 2015, but there is wide variation in this forecast among the panel, ranging from 2010 to 2020. The potential market is vast, in the trillion dollar range.

Smart Robots

Intelligent robots are rapidly being developed that walk and climb stairs, speak with humans, and perform other complex tasks. For example, Asimo, the robot developed by Honda Corporation, works as a receptionist helping visitors find their party, introducing speakers at conferences, and serving tea. Rodney Brooks, who heads a research lab at Massachusetts Institute of Technology (MIT), thinks robots are now at roughly the same stage of development that PCs were in the 1970s.

This progress is largely the result of determination by the Japanese and Korean governments. They know that the proportion of young to older people is declining precipitously, opening a great market for skilled help. There are five workers for every senior citizen today in the industrialized nations, but by 2020 the ratio will decrease to three to one. In Japan, it will be two to one. That's why Japan has funded research in Artificial Intelligence (AI) and robotics, helping Honda, Sony, and Toyota make huge inroads. Toyota plans to replace all auto workers with robots, reducing costs to match those of China. South Korea has organized a task force of 30 corporations to develop a mass market for "national robots" and expect to introduce affordably priced versions soon.

Why are Japanese car makers so keen on robots? Like cars, robots are complex machines costing thousands of dollars, needing check ups and repairs, and the demographic trend toward fewer young people opens a great new market. Honda ran a great ad showing Asimo standing politely and waving to the camera while in the midst of a family on the steps of their home, surrounded by the parents, the children, and the family dog. This is brilliant product positioning.

Collectively, the Japanese and Koreans could easily sell millions of robots to serve important roles in industrial work, home services, healthcare, military, and leisure activities by their target date of 2010. This could breach the critical boundary between humans and machine in ways unimaginable today. A Japanese executive envisions "a robot for every member of the family," and a Korean government official said, "My goal is to put a robot in every home by 2010."

Strong support also comes from the U.S. military. The Pentagon is applying this technology to battlefield use as robotic tanks, land cruisers, and aircraft. The new F-35 Joint Strike Fighter is designed to fly without a pilot. And robot soldiers are in the works. By 2015, the Pentagon plans to have one-third of its military vehicles unmanned.

This is a tough challenge, of course. Smart robots require acute sensory devices, speech recognition, navigation, and AI capabilities that are as yet beyond the state of the art. To maintain natural communication with people, robots have to produce complex facial expressions to indicate human emotions that are difficult to mimic.

Possible solutions to these difficult requirements can be seen in various inventive prototypes. "Watson" is a program built at Northwestern University to interpret emotions by noting the pitch, volume, tone, and tempo of a person's talk. A Japanese robot, "Repliee," has rubber skin, flutters its eyelids, breathes, and moves naturally. Korean scientists have introduced "Ever-1," a female robot with silicon skin that feels human and a face that interprets emotions and responds in kind. Nitta Corporation is planning a five-fingered robot hand. Electronic robotic skin with thousands of tactile sensors has been developed to provide a sense of touch.

More exotic designs abound. The biggest American robot maker, iRobot, is experimenting with insect-like robots and self-assembling robots that reorganize into different shapes to suit conditions. An EcoBot has been developed that powers itself by catching houseflies for energy.

The potential applications are enormous. Healthcare is gearing up to use robots for monitoring millions of patients, taking temperatures, and drawing blood. Robots will play a personal role as companions for the elderly, providing comfort, reminding them to take medications, and alerting authorities if needed. Some believe that robots could pick crops and do other farm work, and in fact, researchers in Denmark have developed a robot that locates and destroys weeds. Robots are now being used to fight fires, handle hazardous materials, work in tunnels, descend into oil wells to find and repair leaks, and perform tasks four miles beneath the surface of the sea. NASA used its robot "Dextre" to service the Hubble telescope, and MIT is working with NASA to develop "RoboNaut," envisioned as an astronaut's assistant.

It's going to be interesting living with robots. I recall speaking with a "virtual robot" named Julie when calling Amtrack. After making train reservations, getting time schedules, and paying her with a credit card, Julie pleasantly asked if I needed anything else? Knowing this was just a software program, I could have simply hung up. Instead, I found myself blubbering sweet nothings to Julie and thanking her for the help. This shows how we love to humanize robots, attributing intelligence, personality, and emotion to them. One of the most popular robots in Japan is a small harp seal who befriends aging seniors, much the people become attached to pets. Roomba, the popular robot that vacuums floors, was described by the *Washington Post* as follows:

> "Women all over America are cooing over their Roombas. They swoon when it hides under the couch and plays peekaboo. When it retires to its nest and settles in for a nap, its power button pulsing like a beating heart, they swear they can hear it breathe. It's robot love."

Our panel of experts is more than 60% confident that smart robots will be used in 30% of homes and offices in industrialized nations by 2022. This will create a huge new industry worth about $450 billion in the U.S. alone.

Smart Sensors

Smart wireless sensors provide the means to control factory operations, monitor environmental conditions, collect military intelligence, detect system failures, and track movements of people and things. Experts say that the MEMS technology used in sensors is advancing quickly, offering the prospect of being able to manage huge systems circling the globe.

The U.S. Army is testing a wireless health-status system that monitors a soldier's vital signs, location, sleep, need for water, wounds, and other critical indicators. The U.S. Marine Corps is combining acoustic sensors, video and infrared cameras, and seismic sensors to create impenetrable airspace by acting as "the eyes and ears of the Marine Corps." Researchers at Cornell have developed sensors no wider than a human hair that can detect light, pressure, temperature, and chemicals, and may possibly be implanted in the body to monitor critical health functions.

Radio-frequency Identification (RFID) tags are growing in use and becoming smaller and more powerful. Hitachi produces RFID chips measuring 0.05 × 0.05 millimeters. Intel Research is producing sensors with complete computers that contain a 16-bit microprocessor, 8K of flash storage, and 256 bytes of random-access memory; just like the cheap, no-battery features of RFID tags, it uses a small antenna to send and receive data and is powered by signals sent from tag readers.

TechCast tentatively expects these devices to reach mainstream global usage by about 2011, with a total market value of $30 billion.

THE COSTS OF PROSPERITY

Reviewing these forecasts strategically using the bubble chart below highlights the huge advances in high-tech industrialization that are coming during 2012 to 2022. Economic growth is likely to be powered by mass customization, exotic materials, nanotech everywhere, intelligent networks of sensors and micromachines, smart robotic helpers, modular home building systems, and other breakthroughs that may surprise us. The powerful knowledge represented by these technologies will in turn encourage industrialization to spread around the globe.

The costs may be as great as the benefits, however. The costs of heightened demands for scarce resources like energy, mounting environmental stress, intensifying competition in world markets, and outbreaks of cultural conflict may also be great. As economists are fond of noting, there is no free lunch. If we are not forthcoming in managing these growth pains, it may take the occasional failure of a nation or region to spur action.

It's interesting that the forecasts for industrialization covered in this chapter coincide with those for sustainability in the previous chapter. It could be an artifact of our method, of course. But I think this means there is a general consensus among our experts that the intelligent technologies driving high-tech industrialization are likely to be accompanied by advances in energy and environmental sustainability. The next five to ten years will prove decisive, therefore. Moving

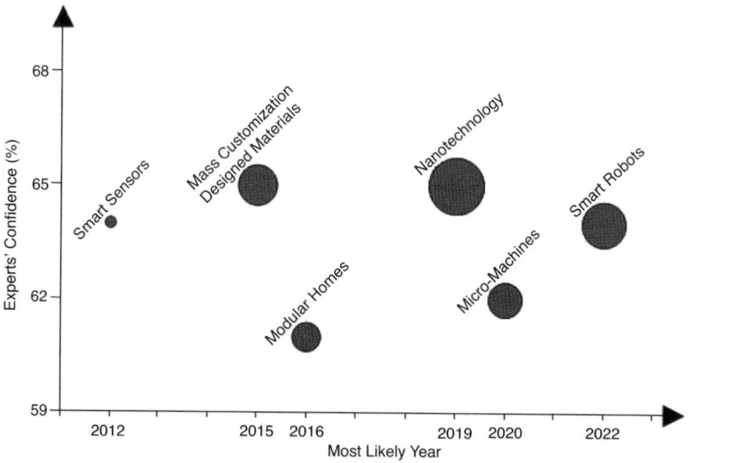

Figure 3.1 Manufacturing & Robotics

toward a fully industrialized world that is also sustainable is an enormous challenge that will test us for decades. It will prove a constant struggle between building a globally integrated industrial system and containing the economic, ecological, and cultural fallout of this historic transition – a worrisome world of plenty.

We should not allow the enormity of this task to blind us to the equally great magnitude of what can be accomplished. Within the brief space of 200 years or so, the world has moved from the poverty of medieval society to imminent mastery over the physical world. The ecological, social, and political barriers will be formidable, as we have cautioned. But the tools for creating a world of abundance are at hand.

Notes

1 Ray Smith, "Our Vision of the Information Superhighway," in *The Infinite Resource* (San Francisco: Jossey-Bass 1996).
2 Thomas L. Friedman, *The World Is Flat: A Brief History of the Twenty-first Century* (NY: Farrar Straus & Giroux, 2007).
3 Joseph Pine and Stan Davis, *Mass Customization: The New Frontier in Business Competition* (Cambridge, Massachusetts: Harvard Business School Press, 1999).
4 *Nanotech Report* (2003).

CHAPTER 4

Society Moves Online: The Transforming Power of Information Technology and E-Commerce

From the first cave drawings, to ancient hieroglyphics, Roman texts, Guttenberg's printing press, telephones, radio, TV, and today's computers, information systems have steadily grown more powerful, and they have changed the world. Think of the striking break between today's world of the Internet and the previous world of room-sized computers, punch cards, and the AT&T monopoly.

Now brace yourself, for the culmination of this awesome power is imminent. Just as vinyl records gave way to CDs and DVDs, all manner of online entertainment will soon flood large wall monitors in our homes and offices with hi-definition sound and video. E-commerce will dominate markets and integrate them into a global economic system. Wireless connections are uniting PCs, laptops, smart phones, iPods, and TVs into seamless networks of digital convergence. Billions of people around the globe are hooking up daily to participative Web 2.0 sites like YouTube, creating a virtual culture of augmented awareness that we all share.

All this is based on existing technologies that are now transforming business and other sectors of society. Threadbare is a web-based clothing company that solicits design ideas on shirts and slacks from millions of customers across the globe, and then gets their opinions on which to produce. "We have a completely different way of doing business," said Jeffrey Kalmikoff, the chief creative officer. "It's community based. You completely erase R&D and a lot of the risk because we just make what people want."[1]

Even more powerful systems lie just ahead. As broadband and wireless provide ever larger bandwidth, computer power continues to rise exponentially, and artificial intelligence (AI) becomes human-like, all this capability will soon allow almost any social function to be performed online. Government services, shopping, education, healthcare, and even religious ceremonies are likely to be conducted *via* the web as realistically as social contact itself. Optical computers, operating with light waves instead of electrons, are likely to arrive in a decade, and biocomputers that store information in the bonds between atoms will follow soon. The ultimate computer – based on the strange laws of quantum mechanics – is in its early stages of development, with the potential for infinite computer power.

Over the long term of 20 years or so, these breakthroughs should increase computer power as much as a billion-fold. Couple this with the graphic feel of intelligent, virtual reality that is starting to enter mainstream use, and it becomes clear we are being transported into nothing less than another realm of consciousness.

HIGHLIGHTS OF IT AND E-COMMERCE

The TechCast Project includes 14 different aspects of IT and 11 aspects of e-commerce – far too much to cover here. Below are highlights of this work focusing on ten of the most significant breakthroughs, organized roughly in the order in which they are expected to arrive. You can always get more details at www.TechCast.org, including references to the facts, quotes, and other items cited here.

Summary of IT and E-Commerce Forecasts

Biometrics Growing at 30 to 40% per year, fingerprints, iris patterns, voice, and other unique bodily features offer the hope of improving security and convenience.

Wireless WiFi, WiMax, EVDO, and Ultrawideband are all appearing in rapid succession to liberate us from that tangle of lines beneath your desk.

Web 2.0 Wikis, blogs, social networks, prediction markets, and countless formats yet to be developed are transforming the web into a bottom-up, participative system.

Entertainment-On-Demand Video is now following music's lead online, making all entertainment available on portable devices and in digital homes.

Global Access IT firms have suddenly discovered the poor, converting the burden of assisting developing nations into huge potential markets.

Artificial Intelligence (AI) With computer power increasing to match that of the brain, routine human thought is expected to be automated by 2020.

Optical Computers Computing at the speed of light is coming along nicely and offers the prospect for replacing silicon chips in ten years or so.
Virtual Reality (VR) The ultimate dream of using IT to transport our senses into space, travel to the bottom of the oceans, or experience your sexual fantasies is almost here.
Quantum Computers Manipulating information at the level of individual atoms may be the ultimate challenge, but it's coming.
Thought Power With researchers capturing brain waves in skull caps, sheer thoughts can control machines and communicate at a distance.

Obviously, the field is evolving constantly, so we are certain to be surprised by newer developments. Some technologies will not work out, new ones may appear, and many details will change. As noted in Chapter 1, however, the broad features of information technology (IT) are likely to emerge generally along these lines.

Biometrics

How to keep ahead of viruses, spam, hackers, and identity theft? Biometrics offers a possible solution by identifying people through their physical features. Fingerprints, hand geometry, the iris, voice, and facial features are all being used, although fingerprints make up 67% of applications. Biometrics is also more convenient because it helps increase security, speed checkout lines, improve customer convenience, and lower costs. Fingerprint scanners costing about $100 can lock your PC or cell phone. Surveys find 82% of Americans support the use of biometrics on passports, 75% on driver's licenses, and 73% on Social Security cards. One expert said "Biometrics is a killer application," and another "The market is set to explode."

Fingerprint scanners are in the works that are simply printed using inkjet techniques and should cost about $2; they are considered 99% accurate because they combine information from the tissue of the finger and the fingerprints themselves. Facial recognition can identify a person even if he has grown long hair. Voice verification is used to authenticate phone callers and credit card users.

Applications abound. The U.S. Federal Government is upgrading its passport and visa system and its employee ID cards to biometrics; by 2010, it intends to build a worldwide biometric information system for the Department of Defense. The U.K. is creating a National Biometrics Identity Register for its 60 million citizens, and the E.U. is issuing biometric passports. In banking, Suruga Bank launched Japan's first biometric deposit accounts in 2004. Biometric drivers' licenses are being used, and schools use biometrics to keep track of students.

Biometrics is not infallible because the human body is infinitely variable. Three percent of people lack readable fingerprints, and 7% have eye pigmentation that interferes with iris scans. But for matters requiring greater security, "multimodal" systems combining two or more techniques are almost foolproof. Yes, someone could possibly use a copy of your fingerprints to breach security, but could they do this with your iris, handprint, voice, or facial features as well?

With the field growing at 30 to 40% per year, it is commonly thought that fingerprints will replace credit cards very soon. TechCast experts estimate that most security systems will use biometrics about 2010 +/– 2 years.

Wireless

Advances in the power and speed of wireless communications are so rapid they are truly revolutionary, with generations of WiFi, WiMax, EVDO, and Ultrawideband following one another in quick succession. The 2nd generation – WiMax – runs at over 70 Mbs, covers a large city, and could soon match other forms of broadband in cost. Sprint is building a new WiMax network that will boost cell phone speeds to match those of DSL and cable. EVDO may surpass WiMax because it can operate over cell networks to provide seamless coverage. Verizon thinks "EVDO could jumpstart the industry all over again." And Ultrawideband operates at 1 Gbs.

Wireless is plagued by clashing standards, spotty coverage, small screens, limited storage, and security problems. Even though the field has been slow in resolving these issues, the desire for mobility is so fierce that the entire world is going wireless, with 60% increases per year. "One-hundred percent cell phone use" has been reached in the E.U. – as many cell phones as people. Americans are a bit slower, but a crossover point occurred about 2005 when more Americans used wireless than wireline phones. Almost all laptops use wireless now, and Sony is putting it in every TV and PC it sells. Robert Kahn, one of the creators of the Internet, said "The world is going wireless," and an International Data Corporation (IDC) researcher called it "The second coming of the Internet."

Behind this rapid growth is an intriguing wave of technological breakthroughs. A new version of wireless – xMax – uses single-cycle modulation to operate with 1,000 times less power than previous technologies. A local network using xMax could run for years on a watch battery.

Some wireless technologies operate across the radio spectrum at low power to avoid intruding on other channels, with the transmissions aggregated into coherent messages at the receiver. Others use "cognitive radio" designs that search for unused spectrum. To overcome gaps between different wireless services, some carriers are using mesh networks that automatically switch to the best service available. Nokia is developing "adaptive beam forming" antennas that focus signals at the mobile device, rather than spraying it in all directions. The result of this astonishing burst of innovation is to convert the electromagnetic spectrum from a scarce resource with limited capacity into an abundant resource of almost infinite capacity.

In Europe, the entire wireless market (WiFi, cell phones, etc.) was $110 billion in 2005 and is expected to reach $1 trillion by 2010. This suggests that potential worldwide demand may be worth several trillion dollars. TechCast estimates wireless will dominate communications by about 2009. Wireless experts think it will reach the astonishing transmission speed of 10 Gbs, able to download movies in seconds, if not in real-time.

Web 2.0

A "participative" Web 2.0 seems to be replacing the original Web 1.0, where content was provided by site managers. Now, bloggers use sites like Technorati to distill the opinions of their 50 million fellow bloggers into the latest buzz. Tagging systems, like del.icio.us and RSS, are creating a more powerful form of search that uses the judgment of millions to identify anything of interest. Open-source software spontaneously designed by volunteers is growing steadily because it is cheaper and more reliable than hierarchy-built systems. The huge popularity of Web 2.0 sites portends the shift from a top-down Internet to a bottom-up Internet in which the knowledge of ordinary people is harvested to create collaborative wisdom and shared experiences. "Power is shifting to the individual and away from giant companies," said the chairman of techcrunch.com. Consider a few of the more prominent developments:

Web 2.0 Applications

Wikipedia This Web 2.0 pioneer pools the knowledge of 65,000 contributors to produce 2 million articles that are as accurate as the Encyclopedia Britannica's 100,000 articles. It has inspired more than 200 similar projects in other languages.

Social Networks YouTube, FlickR, MySpace, Facebook, and about 300 other social networking sites are booming. YouTube shows 100 million videos each day. Cyworld has garnered 15 million Koreans into its fold, about one-third of the entire nation.

Social Search Google and Yahoo are building systems that collect subjective judgments of people to find the best restaurants, shops, or anything else.

Government The U.S. Patent Office is building a wiki to allow knowledgeable outsiders to review soaring patent applications. NASA is using a wiki to encourage wider involvement in critical decisions to avoid the type of O ring failure that caused an earlier launch explosion. Online social networks are used for disaster relief to avoid clogging phones and to enable people to report problems, request help, and get instructions.

Web-Based Software Linux, Microsoft, Google, and Amazon are fueling a boom in web-based software development, often by small groups working in the open-source mode. Microsoft is creating a web version of its Office suite – Office Live.

Citizen Journalism With the number of camera-equipped cell phones approaching one billion, people are capturing images for the media. Kyle MacRae, CEO of "Scoopt," says "Sooner or later, every news story will be captured by a citizen journalist."

Corporate Applications Blogs, wikis, and other Web 2.0 tools are being used by corporations to form an "Enterprise 2.0." P&G formed online forums to learn about consumer behavior, and Motorola has 4,000 blogs and 3,300 wikis.

Assisted Search ChaCha and Amazon provide human-assisted search, in which people are paid to scour for information. ChaCha employs 30,000 "live guides" and Amazon has 100,000 "Turkworkers." Others like Fotolia and iStockPhoto are using masses of individuals to provide content for sale on their sites – crowdsourcing.

Prediction Markets Yahoo, Google, and Microsoft have employees who contribute their knowledge to identifying new products, estimating sales, etc.

Mass-based movements are prone to dangers of group thinking and reckless fads, of course, and are not appropriate in many cases. The Northwestern University women's soccer team posted lewd photos on a site, causing a furor for which they were suspended. When a blogger posted a guide to picking a bike lock with a Bic pen, the lock maker (Kryptonite) was deluged with 2 million complaints for refunds. It cost the company $10 million out of annual sales of $25 million.

Web 2.0 is made possible because broadband can deliver video and other information-rich content, and sophisticated devices (iPods and smart phones) are able to capture this information conveniently. Gartner named Web 2.0 a transformational technology that will have the greatest impact of all IT over the next ten years. TechCast experts think that Web 2.0 will dominate the Internet in a few years.

Entertainment-On-Demand

When Apple broke the logjam in digital entertainment by introducing the iTunes/iPod system, this single venture unleashed a flood of pent-up demand for online music and video. Sales of music albums and CDs are down 20% since then, while online music sales are up 1,700%. Now BitTorrent, Apple, and Microsoft are organizing the entertainment industry to deliver movies and TV shows online. VuDu is expected to turn TV sets into multiplex theaters viewing 5,000 movies supplied by all major studios; the system caches the film's beginning to start playing immediately while the rest is downloaded; "it's so clever that in hindsight it seems obvious," said one expert. Artists are distributing their own music and videos directly to the public as technology lowers costs. A filmmaker called it "a new golden age of entertainment" and *Newsweek* described it as "the democratization of TV and film." A Forrester analyst said, "Once customers try online entertainment, they'll never go back."

With all electronic devices going digital, the long awaited convergence of content and machines is upon us, ushering in the "digital home." TV shows, streaming video, music, voice, and web content are increasingly shared over wireless networks of TV sets, PCs, cell phones, iPods, DVD players/recorders, TiVo, and more to come. (See "Digital Convergence.") Verizon is converting its old phone network to fiber in order to distribute all these media throughout the home. IPTV (digital TV sent over the Internet) is entering the market and could allow unheard of flexibility in storing and viewing programs. The cost of digital TV declined from $5,000 in 2000 to $600 in 2006. Blu-Ray and DVD discs will soon store hundreds of movies, and Turner Entertainment is using holography to store 1.6 terabytes of movies and video.

Unfortunately, piracy haunts the industry. Youngsters, who make up the bulk of the market, often think it's OK to pirate, causing movie and music company losses to the tune of $3 billion per year. "I don't know how you stop this," said a Silicon Valley entrepreneur. The problem will get worse as personal video recorders, digital TV, TiVo, BitTorrent, and broadband make it easy to pass videos and movies on the web freely. Already, 23 million Americans are downloading TV shows illegally, and half of U.S. homes should soon have DVRs and other systems needed to transmit video files.

The solution now emerging is to make entertainment-on-demand convenient and cheap enough to dissuade people from pirating. Experts agree that huge global markets will allow modest charges to become economically feasible, thereby undercutting the profits of pirates. Michael Eisner, former Chairman of Disney, said "If we don't offer products in a timely manner, the pirates will." In late 2007, all were amazed when Amazon, Apple, and others dropped copyright restrictions on their music stores, enabling buyers to treat music purchased for $1 as their own property. It would have been unthinkable just a few years ago.[2]

With continued technical improvements and good business practices, almost all entertainment should be available online soon at reasonable prices. TechCast estimates 30% of all entertainment to be provided on demand by 2009 or so.

Global Access

The challenge of bringing modern communications media to poor nations is enormous. Before the recent boom in global IT, two-thirds of the world's poor had never made a phone call. Poor nations averaged two phones per 100 citizens, and only 3% of their citizens had Internet access. Even in the U.S., only 60% of Americans own a PC. English is becoming common for business and other public uses, but local languages restrict cross-border phone calls, TV, and the Internet.

Globalization is rapidly changing all that as IT relentlessly wires the world. Cell phones, WiFi, and satellites are especially promising because they require less infrastructure. The World Bank estimates half of the world had wireline phones in 2005 and 77% had access to wireless phones. Cell-phone use has multiplied by a factor of 70 over the last decade (about 50 %/year). China alone has 200 million cell-phone users, the largest number on Earth. India went from 1.6 million cell phones in 2000 to 125 million in 2006 and expects to have 500 million by 2010.

And automatic, real-time language translation is expected to be common about 2010 to facilitate cross-border communications.

This sea change in outlook has shifted our view of the world's poor from a burden to a huge new market. MIT, AMD, Dell, H-P, Carnegie Mellon, and Intel are among the prominent entrepreneurs providing PCs for $100 or so to jump start poor nations into the Information Age. A single wireless phone or computer is often shared by an entire village. For instance, in Peru "Internet cabinas" provide Internet access for a small fee, and India is installing "Internet kiosks" in 600,000 villages by 2010. "It is breathing economic life into villages," said an Indian businessman.

Instead of a liability, then, the poor actually represent a huge potential market for inexpensive goods. C.K. Prahalad, a leading business professor, said "The world's four billion poor should be considered the biggest source of growth left."[3] The number of people with incomes of $10,000 per year is expected to double by 2015, and we could see 4 billion to 6 billion global consumers by 2030 or so. Forrester Research estimates the global market for e-commerce should reach $6.8 trillion. Considering all factors, our experts think 30% of the world population will have access to most media by 2016.

Artificial Intelligence (AI)

Since the time Alan Turing defined the "Turing Test" to decide if computers could become indistinguishable from humans, AI has proven both as attractive and as elusive as the Holy Grail. Various forms of AI have beaten world champion chess masters, chatted with humans, guided robots, and most recently steered vehicles through the DARPA obstacle courses. Yet ordinary machines are dumber than children, and many experts think AI has failed its promise.

Perhaps the disappointment with AI lies in the extravagant definition implied in Turing's famous test. Rather than computer intelligence becoming identical to that of humans, a more realistic test would focus on when AI can automate the routine mental work now occupying so much human time. Think of it as "AI that is good enough to work," "weak AI," or "human-level AI." Consider some signs of serious progress:

Developments in Artificial Intelligence

Cambridge University has replaced the attendant at its main entrance with a robotic head that gives visitors information.

Ray Kurzweil notes AI is used now for automatic control of spacecraft, robotic telescopes, analyzing medical results, business and finance, manufacturing, etc. His studies extrapolate the growth of computer power to estimate that a $1,000 PC will match the power of the human brain (10 quadrillion calculations/sec) at about 2020.

All-purpose robots, or androids, are rapidly gaining capabilities. The Japanese and Koreans expect to be placing robots in homes by about 2010, and TechCast finds they will be used in 30% of homes by 2020. (See "Smart Robots.")

Face recognition software has improved by a factor of ten and now exceeds human abilities. The improved algorithms used can even distinguish between identical twins.

Intelligent Systems are becoming smarter. Cyc has been gathering millions of common sense facts and is now good enough to be used by governments and corporations. CogVis learns to play "Scissors, paper, rock" by observing human players.

Speech Recognition is being used by most organizations to replace those maddening touch-tone call centers with smart systems that improve customer service and recover their investment in a year or two.

Five Robotic Vehicles completed the Grand Challenge competition sponsored by Defense Advanced Research Projects Agency (DARPA) for navigating a 132-mile stretch of desert unassisted by humans.

Computer Games are using AI to intelligently guide human action figures in games such as Sims, Metal Gear Solid, Unreal Tournament, and Halo.

IBM's "autonomic computing" program allows servers to reconfigure themselves to accomplish goals, just as we rely on an autonomic nervous system to regulate bodily functions. In a similar vein, Norton provides software to optimize computer performance, fix mistakes, and perform other tasks without user intervention.

DARPA is developing a hypersmart computer that can maintain itself, assess its performance, make adaptive changes, and respond to different situations.

The Department of Energy is creating an intelligent computer that can infer intent, remember prior experiences, analyze problems, and make decisions.

IBM and Microsoft have speech recognition programs expected to improve accuracy greatly by 2010. Dragon, the leading maker of speech recognition software, claims their new systems provide 99% accuracy without training.

MIT is working on Project Oxygen, which features a voice-machine interface. The Project Director, Rodney Brooks, said "I wanted a machine that will look you in the eye, let you ask questions in casual English, and answer them the same way."

IBM is developing a working model of the human brain. Using IBM's Blue Gene, the most powerful supercomputer in the world, the model will simulate a 2-millimeter chunk of the neocortex containing 60,000 neurons. Eight thousand individual processors will be working in parallel, each simulating a few neurons.

Google is providing cross-language search, in which a query in one language can retrieve results in another by providing automatic translation in real-time. The Google VP of Engineering said the company intends "to break the language barrier."

This is just a sample of the talent being poured into AI because of its enormous possibilities. BCC Corporation estimates total AI sales grew from $1 billion in 1993, to $12 billion in 2002, to $21 billion in 2007. Our experts are 60% confident that these various forms of AI will replace 30% of routine mental tasks about 2020 +/– 5 years, producing a U.S. market of $600 billion.

Optical Computers

As silicon chips approach the limit where circuits are only a few atoms wide, computing with light is one of the promising new technologies that could take us beyond Moore's Law. The head of optical research at Bell Labs said "The last 30 years were based on electronics; the next 30 years will be based on photonics."

With information traveling at the speed of light rather than the speed of electricity, optical computers are expected to operate 1,000 times faster. It is especially promising that a large part of the discovery work is completed. High-speed fiber optic cables carry information on light signals, lasers are used to transmit and modulate information *via* light waves, CDs and DVDs use lasers to store and read information in the discs.

To get a sense of the potential, a single optical fiber can carry thousands of channels simultaneously using wavelength division multiplexing (WDM) in which different wave lengths of light (different colors) flow down the fiber in parallel and pass through one another without interference. Lucent has developed an 80-channel fiber and claims 1,000 channels are on the horizon. Ultimately, one fiber should be able to transmit 200 Tbs, equivalent to the entire Library of Congress, in one second. Hundreds of companies are working away in a virtual "Photon Valley" to develop the heart of the machine – a processor that computes using light. Some developments:

Developments in Photonics

MIT researchers have developed a photonic chip that could enter the market in three to five years and could lead to complete optical computers by 2014.

Intel demonstrated a silicon optical modulator operating at 30 Gbs, which could create optical processors far exceeding the power of silicon chips.

An Israeli firm has developed an optical chip, and the CEO thinks "This could become mainstream in ten years."

IBM has developed a silicon chip with nanotubes to emit light, bridging the divide between electricity and optics. The company has also developed methods for storing light by having the signals circle in a resonator, similar to computer memory.

The U.S. Naval Research Lab is controlling light signals with magnets, creating switches that can be used for optical computers.

A Japanese firm developed an optical disk that uses holograms to store 1 terabyte of data in a CD-size disk.

Taming light is difficult, but optical computers seem to be further along than biocomputers and quantum computers, the main competitors for the second generation beyond silicon. A Nortel executive said "An Optical Internet will be the economic engine of the future. It's the most rapidly improving technology in history."[4] TechCast estimates they are likely to enter the market about 2015.

Virtual Reality

The ability to immerse oneself in an artificial environment that simulates all the sensory experiences of commerce, warfare, education, medicine, architecture, entertainment, and space would be a dream come true.

One techie called it "The manifest destiny of computers." VR rooms have been around since the early '90s, although they are expensive and clunky. But governments, the military, and industry have been investing big sums, and we are beginning to see improvements:

Advances in Virtual Reality

Stereoscopic imaging has blossomed to allow true three-dimensional viewing with the naked eye. For instance, some versions of the *Spider Man* and *Polar Express* movies used 3D. Sharp has sold three million 3D cell phones in Japan, and Toshiba is bringing out a 3D TV set. Movie executives see 3D as "a stereo renaissance."

Internet 2 offers "teleimmersion," which allows people to walk around a virtual object and to hold virtual meetings with three-dimensional images.

The U.S. Department of Defense (DoD) is developing a system to produce six-foot holographic images in color.

Japanese researchers have introduced "aroma seats" in movie theaters that waft various scents depicted on screen.

Sony is developing ultrasound systems that target parts of the brain to create sensory experiences of sight, sound, smell, taste, and feeling.

Cisco is introducing high quality video conferencing systems at low prices. An analyst said "Good systems can provide 90% of the experience at 25% of the cost of travel. There will be a market for this."

Computer games now appear in 3D and are vividly realistic. A reviewer of Microsoft's Xbox 360 said "I didn't appreciate war until playing this game."

Online avatars are so sophisticated they can express joy, love, and laughter; walk, run, and sit; talk, date, buy, and sell. Second Life, Yahoo, and AOL are flooded with avatars relating to other avatars.

Just as flight simulators are indispensable for training pilots, these capabilities are being used to train physicians, assist surgeons in operations, test the design of buildings and vehicles, allow researchers to analyze data insightfully, create interactive forms of entertainment, and hold virtual meetings with full visual cues. And the technology is improving rapidly as broadband, wireless, speech recognition, AI, and more powerful computers go mainstream. (See "Intelligent Interface").

The following prominent applications would have been unthinkable a few years ago:

Prominent Applications in Virtual Reality

The UCLA Virtual Reality Lab recreates ancient Rome with 3D images of temples and monuments that users can walk around. The lab head called it "A kind of time machine."

Courtrooms are starting to use "Instant Scene Modeler" to create 3D images that allow analysis from different angles and make it possible to zoom in on details.

Japan is planning to introduce "virtual reality" TV in 2020 that includes 3D images, touch, and smell.

Planetariums have begun to give people the experience of flying through space.

The E.U. is planning to augment the experiences of tourists to Pompeii with virtual images of ancient Romans going about their daily tasks.

Telesurgery can be performed miles away using VR to assist the surgeon.

Virtualized reality allows sports fans to view the action from any location, like riding a football from the quarterback to the receiver.

Soldiers are being trained using computer games and complete simulated battles.

Data analysis can visualize hidden relationships in complex data insightfully using VR. For instance, VR is used to make more accurate decisions in financial markets, and scientists can observe the effects of different theories in vivid data displays.

The Waldorf Astoria Hotel in New York City has been using a 4 × 7 foot video conferencing system for years, and it is so convenient that it is always booked. "It feels like you are all sitting in the same meeting," said one user.

All this is coming together to form a complete "virtual universe" or "metaverse." Virtual worlds, like Second Life, have millions of users and are growing rapidly. "There" is a multimedia website featuring 3D computer-generated environments populated with avatars that interact with users. Google Earth has spawned a flood of mashups, and the thrill of zooming in anywhere from space make it extremely popular, with 250 million users as of 2007. *World of Warcraft* has eight million users, and *Lineage II* has 14 million.

VR is growing 33% per year, and experts think the Internet will be dominated by VR about 2015. Gartner thinks 80% of *Fortune* 500 firms will use these systems by 2011. Our experts agree VR is likely to enter homes and offices about 2016, allowing virtual meetings with life-like 3D images, intimate conversations with distant friends, and almost any type of social transaction. (See "TeleLiving").

Quantum Computers

The strange behavior of matter at the quantum level offers another possible successor to silicon in which information is stored in individual electrons. The basic unit of quantum computing is a "qubit" – an electron spinning either clockwise or counterclockwise, representing a 0 or a 1. Because electrons can coexist in two places simultaneously, this feature of "superimposition" allows a single electron to carry two qubits, two electrons can produce four qubits, three electrons eight, and 20 electrons could perform a million computations. This exponential growth raises the hope of infinite processing power. A quantum computer could easily complete in seconds a task that would take a silicon computer billions of years.

This vast power enables cracking even the most sophisticated encryption codes in a flash. Conversely, it is impossible to observe a quantum state without altering it, so quantum cryptography can detect eavesdropping. Despite the daunting challenge involved, the first research prototypes of quantum encryption are now protecting sensitive data at Harvard University, the National Security Agency (NSA), and the Federal Reserve.

Some scientists caution that the indeterminate nature of quantum physics is troublesome. If quantum computing relies on the ability of atoms to assume two states simultaneously, this uncertainty may produce errors. The data stored in qubits may be unstable and only hold a second or so before deteriorating into "decoherence."

The fascinating thing about quantum computing is that it leads to the possibility of teleportation, as in "Beam me up Scotty." Two or more atoms can become "entangled" such that a change in one instantaneously produces a like change in the other across great distance. Entanglement has been show to transport information more than 1 km thus far, and physicist David Darling thinks "Teleportation is going to play a major role in our future."[5]

With such intriguing prospects, many research institutions are investing large sums to advance the field:

Advances in Quantum Computing

D-Wave Systems in Canada claims to have built the world's first quantum computer. It operates using 16 qubits and is expected to have 1,000 qubits in 2008, enough to do massive tasks. Some scientists doubt if it uses true quantum computing methods, but Seth Lloyd at MIT said "This work is potentially solid."

IBM has demonstrated the ability to store one bit on a single atom; in comparison, hard drives use one million atoms to store one bit of information. Using this technology, an entire supercomputer could be the size of a speck of dust.

Quantum memories have been built using "spintronics" to control the spin of electrons: right spin is a "0" and left is a "1." The first quantum memory that used spintronics increased data density 1,000 times.

Scientists at Penn State have created a 3D array that holds hundreds of atoms. It uses lasers to create an "optical lattice" that contains 250 atoms precisely at fixed locations where they can be manipulated. A device with 250 atoms is more than enough to perform vast computational operations.

An "ion trap" has been constructed to control individual electrons in switches (like a silicon chip), laying the foundation for mass production of quantum computers.

Georgia Tech researchers were first to demonstrate that it is possible to transfer information from atoms to light particles, or photons.

A variety of researchers have controlled superposition, photographed electrons, flipped the spin of atoms, teleported information, and entangled up to five photons.

Looking at the distribution of TechCast expert opinion, some are doubtful about the heroic task of computing at the quantum level. There is a strong cluster of confidence levels in the 50 to 80% range, however, and our best forecast is that quantum computers are likely to become available about 2021 +/– 5 years.

Thought Power

Experiments have shown that individuals can direct their thoughts through brain signals to communicate with computers, robots, and other people. These are primitive systems as yet, but they represent a beachhead in the new area of brain-computer interfaces that is now being explored seriously. As the nervous system is better understood and the technology improves, we could extend the power of sheer thought to all manner of devices, gain mastery over our bodies, and even communicate with each other silently at great distances.

Scientists have used skull caps containing sensors to capture electromagnetic brain waves for controlling various devices. The system learns and adjusts to the individual, and has shown that paralyzed people can write at proficient speeds. Others are implanting electrodes directly into the brain. "EEG is fine for moving a cursor, but you need more control," said one scientist. "The deeper you go into the brain, the better the signal." Paralyzed patients with imbedded electrodes have learned to control TV, read email, use the Internet, and write letters.

Brain implants should become increasingly feasible with miniaturization, and wireless transmitters will allow signals to be sent over great distances. But the drawbacks are not understood. It is easy to image infections in the central nervous system, loss of control as devices malfunction, and a host of other dangers. But the potential is drawing forth fascinating experiments:

Advances in Thought Power

Prof. Kevin Warwick, University of Reading, England, was the first human to implant a computer chip directly into his body. It enabled him to control electronic devices and communicate electronically without words to his wife, who had a similar chip implanted.

The US Food and Drug Administration (FDA) approved an implantable chip the size of a grain of rice for storing medical data under the skin of human patients. Similar chips have been imbedded in one million pets and in about 1,000 people in other nations.

Cyberkinetics Corp. plans to use its system, "BrainGate," to restore function in paralyzed people. The system uses a skull cap with sensors to capture thoughts and direct them to controlling machines and communicating with people.

Artificial arms and legs are being wired directly into nerves so that the user's normal thoughts can control them. (See "Artificial Organs").

The U.S. military is working to help soldiers gain mental control over their thoughts and body so they go without sleep for days, or to stop the flow of blood from wounds.

Security is made easier using brain waves, which are unique for each person, as biometric passwords. A mere thought or touch can serve as a password to access buildings, financial accounts, and the like, or to exchange information by simply touching the recipient.

Microsoft Research is developing a skull cap that uses learning algorithms to raise the accuracy of measuring brain signals to 99%. They expect it to be used for video games, workplace productivity, and to simplify using a PC.

Hitachi is developing a skull cap that measures changes in blood flow. Potential applications include controlling TVs and intelligent autos.

Omneron Corp. is using real-time MRI scans to help patients mentally control pain, anxiety disorders, depression, and the effects of stroke.

This suggests the possibility of "transhuman" capabilities enabled by thought – artificial limbs, electronic eyes, cochlea, and other organs creating "bionic humans" with extraordinary powers. The Director of DARPA envisages warriors using this technology to remotely control aircraft and robots. The potential was envisioned by physicist Freeman Dyson as "Radiopathy" – the ability to implant wireless chips in the brain, connecting people silently with each other and with intelligent machines.[6]

TechCast experts are 60% confident that brain-computer interfaces enabling people to communicate mentally and better control their bodies and distant objects will be commercially available about 2020.

INFORMATION TECHNOLOGY CHANGES EVERYTHING

The bubble chart for e-commerce below highlights how commercial use of the Internet should reach 30% "take-off" levels during the next decade. Most web use today is for casual searching, with paid commerce at minimal levels. For instance, e-tailing, B2B, and other forms of e-commerce run in the 10% range. It's going to be interesting when trillions of dollars are traded over the web in a few years, transforming industries and altering lifestyles. *BusinessWeek* noted: "The Web is the same age color TV was when it turned profitable."

The chart shows that advances in intelligent IT systems should continue their hectic pace, culminating in a second generation of optical, biological, and quantum computing about 2020. The year 2020 seems to be emerging as a pivotal time, therefore, when the IT revolution reaches its zenith. Computer power has increased roughly a billion-fold since mainframes were first introduced, and it is going to increase another billion-fold. For the first time in history, personal

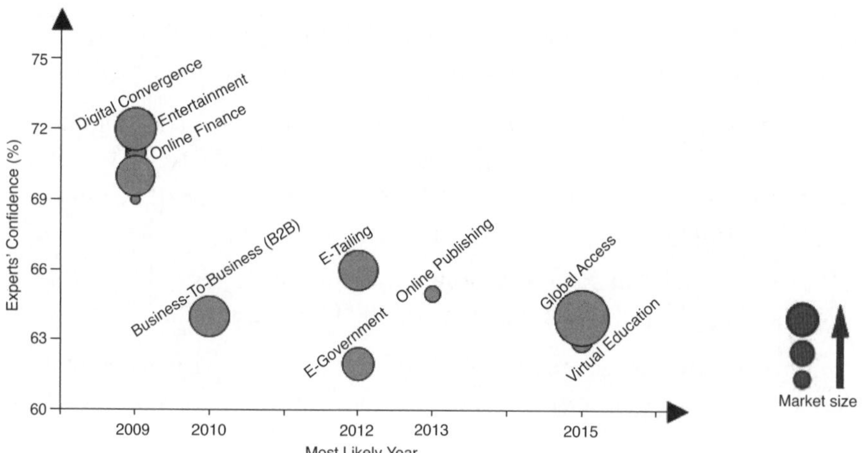

Figure 4.1 E-Commerce Forecasts

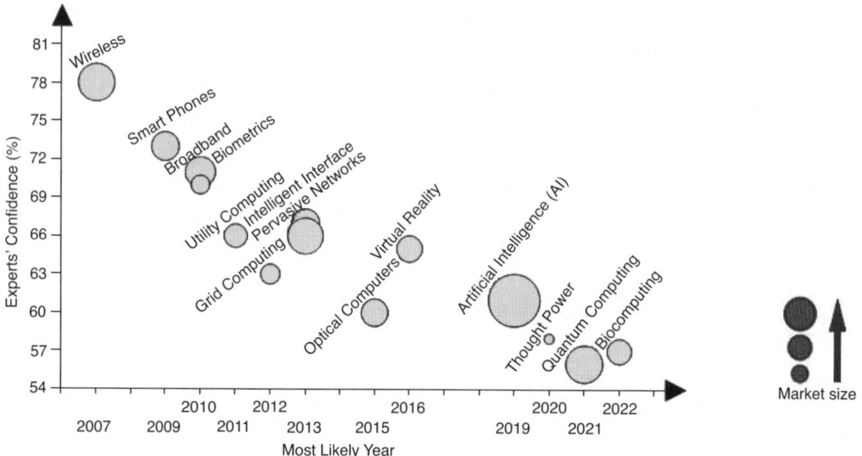

Figure 4.2 IT Forecasts

communications, graphics, video, data flows, and all other forms of information should be used routinely by ordinary people anywhere, with profound consequences. Think of the liberating effects of the printing press alone, and multiply that by thousands or millions.

The Intelligent Interface – TeleLiving

Perhaps the most striking change will be replacing today's "dumb" computer interface with an "intelligent interface." Speech recognition,

AI, and virtual environments are all likely to be widely adopted by 2010 to 2015, producing a "conversational" form of human-machine interaction. If current trends hold, a modest version of the talking computer made famous by Hal in *2001, A Space Odyssey* should soon become a reality. Rather than hunch over a keyboard or mouse, the PC will disappear into a corner while we talk to life-sized virtual persons on large wall monitors. I call it "teleliving" – a conversational human-machine interface that allows a more convenient way to shop, work, educate, entertain, and conduct most other social relationships.[7] This growing trend outlines the great arc of IT's trajectory from the tele***phone***, to tele***vision***, to tele-***living***.

Almost any social transaction – teleworking with colleagues, buying and selling online, taking an educational course of studies, consulting with your physician, or just a casual talk with a distant friend – could soon be conducted in a conversational mode, speaking with life-sized images as comfortably as we now use the telephone and television. It should feel as though virtual people are right there in the same room with you. It could be a boon for "virtual assistants" (intelligent software that helps with ordinary tasks), "virtual tutors," and other useful applications.

The power of this concept can be nicely illustrated with a hypothetical shopping experience. Suppose you want to buy a book. Today you would go to Amazon.com, find your way through a complex homepage containing 20 to 30 items, and then click through successive pages of instruction. Now suppose you were greeted instead by a "virtual salesperson" who knew you by name? After asking what you wanted, the salesperson would find the book, display it with price and shipping, ask for your credit card, and inquire if there were anything else you needed – just like a real salesperson, but one who knows all the merchandise and has infinite patience. You could probably alter its character about as easily as changing wallpaper on your PC. Would you rather be served by Madonna, Rudy Giuliani, or Hillary Clinton? Click here.

A few years ago Bill Gates claimed "The future lies with computers that talk, see, listen, and learn." This view is now supported by other computer industry leaders. Robert McClure of IDC, stated recently "What the graphical user interface was in the '90s, the natural user interface will be in this decade." Sony President Kunitake Ando expects the PC to be more of a "teacher, agent, and guide." Our TechCast expert, Ian Pearson at British Telecom, sees "better interface technology ... and AI." And computer scientist Ray Kurzweil forecasts

"It will be routine to meet in full-immersion virtual reality for business meetings and casual conversations in five to seven years."

Shifting Social Foundations

We are fully aware of the cynicism that persists over the unrealized promises of speech recognition and AI, and we know the intelligent interface will present its own problems. If you think today's dumb computers are frustrating, wait until you find yourself shouting at a virtual robot that fails to grasp what you badly want it to do.

Despite the abundant problems, the social order is now shifting beneath our feet. Just as industrialization shed the claustrophic feudal system for modern governments and corporations, today IT, knowledge, and intelligence are shifting the foundations of society once again. Web 2.0, smart phones, e-commerce, biometrics, virtual reality, and wireless everything are introducing distributed information systems to manage the complexity of modern life, and the social order is following suit. In Chapter 8 we will see how the logic of knowledge differs from that of capital, with profound ramifications for creating bottom-up, quasi-democratic institutions operating in real-time.

The impacts are so profound that they alter the very definition of who we are. As I noted in the introduction, all this computerized intelligence is "automating" the routine aspects of human thought itself, leaving people to do – exactly what? Just as the automation of farming, manufacturing, and services failed to leave us bored, the automation of mental thought is likely to raise the level of human attention again. As Chapter 9 will argue, I think we are likely to discover that humans are best at all those messy, subjective, creative tasks that machines can't do and that increasingly dominate attention. It should be nothing less than an historic shift to a higher level of consciousness needed to manage a more difficult world.

I think it is reasonable to say that the maturing of IT during the next decade or so is likely to change everything. Lifestyles, economics, markets, social institutions, and the very notion of what it means to be human are all being transformed. What could possibly be more fascinating intellectually and more important practically?

Notes

1 "Children of the Web," *BusinessWeek* (July 2, 2007).

2 Rob Pegoraro, "The Sound of Cop Restrictions Crashing," *Washington Post* (May 17, 2007).

3 C.K. Prahalad, *The Fortune at the Bottom of the Pyramid. Eradicating Poverty Through Profits* (Upper Saddle River, NJ: Wharton School Publishing, 2005).
4 BusinessWeek (6/12/00).
5 David Darling, *Teleportation* (NY: Wiley, 2005).
6 Freeman Dyson, *Imagined Worlds* (Harvard University Press, 1997).
7 William E. Halal, "The Intelligent Internet," *The Futurist* (Jan–Feb 2003).

Chapter 5

Mastery Over Life: Promises and Perils of Biogenetics

The advance of medicine has been long and slow. As George Washington lay dying from a throat ailment, his physicians bled him occasionally to "relieve the body of foul humours" – actually draining away the very energy needed to survive. Just a few decades ago, before the discovery of penicillin, we were helpless against wave after wave of tuberculosis, typhoid, plague, smallpox, whopping cough, polio, and other horrible illnesses that swept through society at will. Today, less than 10% of deaths in modern nations are due to infectious disease, while the other 90% are the result of environment, genetics, and lifestyle.[1]

That simple fact sums up the reason medical care is now so complex and expensive and getting more so. Modern medicine is employing advances in genetic engineering, molecular biology, nanotechnology, IT, artificial intelligence (AI), and many other fields to address far more difficult technical problems at the atomic level where life originates. The challenges are so tough that these advances will take roughly two decades to arrive, but they will mark a major leap forward. Just as we are horrified now by the thought of bleeding patients, we are likely to marvel at the clumsiness of today's health practices.

THE MIRACLE OF DNA AND OTHER BREAKTHROUGHS

The summaries below highlight the major breakthroughs we will explore in this chapter. As noted before, references are from www.TechCast.org. Of the eight technologies covered, only telemedicine and artificial organs offer hopes of an impact reasonably soon. All the others depend on converting the discovery of DNA into practical benefits, and that will take hard scientific work and time.

Summary of Medical Forecasts

Telemedicine Electronic medical records, online doctor's visits, robotic surgery, and other uses of IT should curtail exploding costs and improve quality of care.

Artificial Organs An astonishing range of organs can now be replaced with artificial parts, and the replacement of most major organs is likely about 2015.
Child Traits Despite the sensitivities involved, gender is being chosen now by parents, and it's likely that intelligence, height, and other traits will also come under control.
Grown Organs Tissue engineering, stem cells, and even regeneration of limbs are being used experimentally to repair and regrow almost all body parts.
Cancer Cure It may sound wishful, but nanotechnology and other cures offer the hope of restoring normal lives to cancer patients by 2022.
Personal Treatment Understanding DNA allows treatments targeted to each person, improving health and reducing the side-effects of one-size-fits-all medicine.
Genetic Therapy One of the great lures of biotech is eliminating the 5,000 genetic diseases that plague humankind. Progress is slow, but results are likely about 2024.
Life Extension Average life spans are likely to reach 100 in 30 years or so, and they may even exceed the "natural" limit of 120 years.

Telemedicine

Medicine is probably the least computerized industry in the world. The problem is that healthcare is a very complex field, but it is also stalled by institutional resistance. With out-of-control costs and more powerful IT, however, progressive hospitals are embracing online medical records, virtual exams between patient and physician, robotics, and a host of other forms of telemedicine. These advances should enter the mainstream in about a decade to save hundreds of billions of dollars, greatly improve healthcare, and provide more convenient service.

The enormity of the need is seen in the almost total lack of medical IT. Although physicians now often used hand-held PDAs, it is estimated that 90% of all transactions are still conducted by paper and telephone in the U.S., and electronic medical records are only used by 17% of hospitals. "Medicine remains a paper-driven system," said a professor of medical informatics at Stanford. It does not help that hospitals are bureaucratic, insurance usually does not cover telemedicine, physicians resist computers, and patients are fearful about privacy. And poor information systems often discourage use. After automating a medical procedure, a physician complained, "A task that once took three minutes suddenly devoured half an hour."

But the advantages are equally enormous. Some think telemedicine could save as much as one-third of the $1.7 trillion the U.S. spent on healthcare in 2005; this is expected to reach $2 trillion by 2007, or 20% of the entire GDP. Telemedicine could also eliminate 80% of the 98,000 deaths per year caused by medical errors. Many employers like GM and GE only patronize computerized hospitals because they are able to reduce costs by 30% to 40% while improving service. Dell provides its employees an online healthcare system. "Every corporation is considering this," said the CEO of WebMD. Some telemedicine trends:

Trends in Telemedicine

Electronic Records "Point of care" systems allow health workers to access electronic patient records, make notes, prescribe treatment, and order drugs. While the U.S. lags in electronic records, Finland, Norway, and Sweden have largely converted to electronic systems, while France, Germany, and Denmark are in progress now.[2] China's Center for Disease Control has an IT system that allows daily updates from 16,000 hospitals.

Conference Systems "Teleconferences" allow patients to be seen at a distance and enable medical personnel take diagnostic data. Some 32 U.S. states now provide online consultations, and half of patients say they prefer virtual visits. The data shows quality of care remains unchanged.

Intelligent Diagnostic Systems Complete models of patients are being constructed from medical data to diagnose illness. Computerized diagnosis offered by IBM has been shown to be more accurate than diagnoses by physicians.

Robotic/Telesurgery Use of the Da Vinci robotic surgery system has grown from 1,500 operations in 2000 to more than 20,000 today. A robot guided at an Italian hospital performed the first operation online (heart surgery) on a patient in Boston. A woman in France had her gall bladder removed by a surgeon in New York City.

Exemplars Kaiser Permanente is completing a $2 billion web-based system that will help 10,000 medical personnel serve nine million patients at 362 hospitals. Brigham & Women's Hospital system in Boston uses 30,000 workstations to integrate all healthcare for 700,000 outpatients.

The IT Enabled Physician Physician training is now IT intensive, including the use of PDAs, CD-ROM, and patient simulators. "The computer is the physician's black bag of the future," said the dean of the Harvard medical school.

With such great stakes involved, it's easy to see why many think the computerization of healthcare is inevitable. A survey found 70% of hospitals are planning to adopt telemedicine, and almost all patients think it's a good idea. If hospitals don't move quickly, they may find tech-savvy patients demanding it. Forty million Americans now use the Internet to find health data. Telemedicine is growing 30% to 50% per year, and the CEO of Waterford Telemedicine expects it to cover a large part of all healthcare soon. TechCast estimates telemedicine will be used 30% of the time to maintain medical records, order drugs and lab tests, diagnose illness, monitor patients, and other medical work by 2014.

Artificial Organs

An astonishing array of body parts can now be replaced with artificial equivalents: skin, bone, blood vessels, cochlea, heart valves, pacemakers, knees, hips, etc. Using a combination of computer chips, micromachines, tissue engineering, and other new technologies, artificial organs may soon be available to replace major parts of the human body. "Name almost any human disability," wrote *BusinessWeek*, "and there's probably research underway to overcome it."

Medicine has a long history of success in replacing bodily parts. In the U.S. alone, 150,000 knees are replaced by metal joints each year, to say nothing of pacemakers and heart valves. About 90,000 people have electrodes in their heads to control tremors of Parkinson's disease, and 70,000 have artificial cochleas to restore hearing. Pressure to develop new prostheses is intense because organ donors are in such short supply that 100,000 people die in the U.S. each year while on waiting lists, and organ transplants struggle against immune rejection. The field is advancing rapidly as Moore's Law drives IT forward, so almost all bodily parts could in principle be replaceable with artificial counterparts in no more than two decades.

Artificial arms and legs now use computer chips to coordinate movements, are wired into the nervous system so they can be controlled by normal thought, use sensors to provide a sense of touch, and have micromotors to power joints. Some are designed to precisely match the user's limbs, including freckles and hair. An artificial hand can move all fingers and an opposable thumb.

Researchers have even been working to restore sight. Video cameras are installed in eyeglasses, transmitting images to chips in the back of the eye wired to the optic nerve. Lasik surgery can now map the wavefront of the lens and sculpt it precisely to correct imperfections. Another approach implants a corrective lens into the cornea. An opthomologist estimates, "Artificial vision will soon allow blind people to move around freely; within 25 years they could read."

An artificial heart has been approved for temporary use, along with artificial blood. An artificial kidney is being tested that may eliminate the need for dialysis and kidney transplants. Medtronics Corp. is testing an artificial pancreas that delivers insulin to diabetics. Even the brain is being replaced bit by bit – the ultimate prosthesis. Computer chips are being used to augment the hippocampus where memories are stored. Medtronics and other companies are installing "nueromodulators" – coin-sized chips implanted in the brain to control epileptic seizures, depression, migraine, and other disorders. Unlike drugs, there are no side effects.

Some of this seems ghoulish, especially messing with the brain. But who would have thought a few decades ago that we would be transplanting thousands of hearts every year? And it is disconcerting to know that your life hinges on a piece of machinery that could fail. An artificial heart using a small turbine (rather than a pump) has been developed that is small and durable, but it produces no pulse. A friend with an artificial heart valve spoke of the constant terror he felt wondering if that steady thumping in his chest would stop. He also said it was a great blessing, considering the alternative. TechCast estimates that artificial organs will replace the heart, lungs, kidneys, liver, and other major body parts at about 2020 +/– 4 years.

Child Traits

The selection of children's genetic traits is one of those technologies fraught with controversy. Simple characteristics, like gender, can be selected at will now, and it should be possible to select almost all traits as genetic engineering gathers steam. But people hold different views on the morality of this practice, many considering it a form of "playing God" that should be prohibited, while others think it should be left to the discretion of parents.

In the U.K., 80% of the public disapproves of selecting a child's sex, and the practice is regulated because "The social benefits do not outweigh the possible harm." The Council of Europe bans genetic engineering. It does not help when the media inflames this sensitive issue with talk about "designer babies." Other cultures believe selecting a child's traits is not philosophically different from pro-choice abortion, and it could solve serious problems. For instance, children with serious genetic defects, like Down syndrome, are so stigmatized that they are usually aborted. Three-fourths of fertility clinics in the U.S. provide "prenatal screening tools" to detect and abort deformed fetuses. Patients say "We want to prune this from our family tree forever."[3]

Biogenetics is generally leading to the control of DNA characteristics, however, and it's hard to fault parents who think this is best for the child. Even now, sperm sorting can produce a boy or girl with great accuracy, and in vitro embryos can be selected based on sex with 100% accuracy.

The historic trend is that social norms evolve over time, so medical practices considered taboo often become acceptable. After all, it was once considered wrong to open the human body, transplant organs, or conceive children in vitro. Our experts are impressed with the social barriers, however, so they estimate that it will be about 2030 before 30% of parents are likely to alter genetic traits of their children. It will be interesting to see how this dilemma plays out in the years ahead.

Grown Organs

Imagine the benefits of being able to grow living organs in a laboratory from a patient's own cells, producing replacements that are genetically identical. No rejection problems. No organ shortages. And no end to our ability to repair damaged parts of the body. Human skin, bone, and liver tissues are being produced genetically, and the same method is being extended to create entire organs. Stem cells are being used to regenerate damaged organs and cure intractable disease. One neurobiologist calls them "magic seeds." The ethical dilemmas presented by this research are daunting, as are the scientific challenges. However, some scientists claim a veritable body shop of lab-grown organs will wend its way from labs to patients in about 10 to 20 years.

The need is huge. Worldwide, 150,000 people wait for organ transplants, and many illnesses could be cured by engineered tissue. At least 200,000 people suffer from spinal-cord injury. One million have Parkinson's disease, 4.7 million have congestive heart failure, and millions are diabetic. The gap between the number of people waiting and the organs available is increasing by 10% to 15% per year.

The nascent field of tissue engineering is ready for prime time. Lab-grown bone, blood vessels, skin, and other organs are being tested in humans. A living jawbone, nose, bladders, and an ear have been built by growing cells on a scaffold, while livers, breasts, hearts, and fingers are under way. "Bioprinting" techniques are adapting inkjet printers to "print" layers of tissue using "bioink" consisting of live cells to build entire organs. "We can print any desired structure," the researchers claim. Polymers, live cells, nanotubes, and growth hormones are being injected to repair damaged organs or grow new cartilage and bone.

A variety of methods are being developed to grow stem cells without raising ethical concerns about destroying embryos. Researchers have successfully converted skin cells into stem cells, demonstrating the critical ability to produce stem cells from adult cells. Some think it should be possible to convert adult cells into stem cells by switching genes. Other researchers have converted stem cells into brain cells, offering hope for repairing damaged parts of the nervous system. Researchers have discovered the gene that makes stem cells "pluripotent" – able to grow into any cell of the body.

Applications are underway. The U.S. Food and Drug Administration (FDA) approved a trial to inject stem cells into human brains to treat neural disorders. Abbot Labs, Schering, and Genzyme are harvesting stem cells, growing larger quantities, and coaxing them to generate skin, heart muscle, nerve and brain tissue, and entire organs. "Within four to six years, stem cells could become the regimen for patients with damaged hearts," said a scientist.

It may sound like science fiction, but there is a good chance we may even learn to regenerate organs as some animals do. Mice with certain genes have regenerated heart, toes, joints, optic nerves, and other organs, and cells from the mice produced the same effect when placed in normal mice, suggesting there is a genetic solution. Other trials with mice show a protein can regenerate optic nerves after they have been cut. A researcher called these advances "a new chapter in regenerative biology."

There is a fine line between coaxing cells to grow, however, and "errant" cells that could turn cancerous. One study found that stem cells mutate with succeeding generations, increasing the likelihood of abnormalities. A neuroscientist at the U.S. National Institutes of Health noted, "Stem cells may be more cancer-prone."

Organ transplants and other common procedures today were considered equally dangerous not too long ago, so these problems might be overcome in time. Our studies show it is likely that organs will be produced genetically to replace major body parts (kidney, liver, heart, etc.) about 2027.

Cancer Cure

After decades of limited progress, today smarter drugs, nanotechnology, molecular biology, and genetic engineering are producing far more sophisticated treatments that are more effective and safer. "I think we are going to see a revolution in cancer prevention and treatment in just a few years," said a scientist.

Ninety tests are available to detect cancer earlier and more accurately. There were only ten cancer drugs in '95, but more than 400 are in testing now. Two new drugs – Tarceva and Avastin – can shrink tumors 90%, making the disease manageable. A vaccine for pancreatic cancer raised two-year survival rates from 15% to 76%.

The most promising treatments include 60 or so forms of molecular biology and nanotechnology, which are especially useful because these tiny molecules and intelligent devices can kill cancer cells precisely with no side effects. Nanotech agents can be designed to seek out cancer, they are small enough to enter the cells and destroy them, and are safely removed later by the kidneys. Some trends:

Trends in Cancer Treatment

Microbiology RNA molecules 25 to 40 nanometers wide are perfect for identifying cancer and carrying drugs to destroy it. "Biomarkers" can identify early signs of cancer by detecting aberrant gene behavior. Molecules called "dendrimers" are used to deliver drugs precisely to cancer cells. A researcher said "We have not had an efficient system to deliver drugs. This is an incredible accomplishment."

Designer Bugs Bacteria are being designed to seek out cancer, enter the cells, and produce a toxin that destroys them. "Bacteria are the ultimate in smart drugs," said a geneticist.

Smart Nanotubes Silicon nanowires detect cancer and also indicate the cancer type. "These devices distinguish among molecules with near perfect selectivity," said a researcher. Metal-filled or coated nanotubes are able to detect and destroy cancer by delivering drugs or other agents Bundles of nanotubes act as lasers when exposed to radiant energy, destroying cancer cells with bursts of light.

Bioengineering DNA is placed in nanocapsules to deliver genetic therapy for curing the biological cause of cancer. Nanobodies consisting of extremely small proteins have an ability to fight foreign bodies, but are small enough to penetrate tumors and other dangerous cells. Researchers have found mice that are totally resistant to cancer. Injecting their white blood cells into other mice with cancer killed all the cancer cells.

Basic Causes "Cancer stem cells" have been discovered that resist standard chemotherapy and go on to trigger regrowth of tumors. "We hope to destroy the engine responsible for treatment failure and recurrence," said a researcher.

Large Data Bases The U.S. is planning a Human Cancer Project bigger than the Human Genome Project. The National Institute of Health (NIH) will spend $1.35 billion over nine years to identify mutations that cause cancer, which will form a "Cancer Genome Atlas." A researcher said "Knowing [genetic] defects points to the Achilles heel of cancer."

Andy von Eschenbach, head of the U.S. National Cancer Institute, thinks cancer could be eliminated as a cause of death by 2015. Our experts are not quite as optimistic. They think life expectancy of cancer patients is likely to approach normal life spans by 2022, allowing cancer patients to recover and lead full lives.

Personalized Treatment

Like most things in life, people vary enormously in their genetic susceptibility to illness, drugs, and other factors, making one-size-fits-all healthcare often ineffective and at times highly damaging. The generalized approach is only 40% to 50% effective and often produces serious side effects. More than 100,000 people die each year in the U.S. from side effects of drugs, and another two million become seriously ill. Now that the human genome is being analyzed carefully, researchers are moving toward genetic tests to determine these differences and thereby permit precise medical treatments that are more effective and safer. The Director of the MIT Genome Center called it "The framework for the future of medicine." Here are some signs of progress:

Trends in Personalized Medicine

Testing Companies now offer genetic tests for a few hundred dollars to determine predisposition for cystic fibrosis, blood clotting, breast cancer, and other diseases.

Costs Dropping In 2006 it cost $10 to 20 million to sequence a person's DNA, but rapid technical advances are expected to reduce that to $1,000 in a few years. The National Human Genome Institute is developing "nanopore" technology that could slash the cost of mapping DNA in five to ten years. Dr. Craig Venter, who deciphered most of the human genome, launched the world's largest genome sequencing center to make genetic tests commonly available. "Our goal is to do an entire genome analysis in minutes or hours," he said.

Identifying Differences Research is identifying which patients will respond to different cancer drugs. IBM and the Mayo Clinic are developing a system that analyzes a person's medical history and DNA to spot likely illnesses. Other studies show a pronounced interaction between foods and genetic activity. "When you consume food, your genes light up like a Christmas tree," noted a scientist.

Advanced Diagnosis "Breathalyzers" can accurately detect the chemical traces of cancer, TB, diabetes, and other illness. "Give us the chemical fingerprint of a disease, and we can devise a test for it," one scientist said.

Large Data Bases The "Human Variome Project" is a database of all human genetic mutations, allowing rapid diagnosis of disease. And the "Personal Genome Project" is designed to integrate all medical data on each person. The chief of research at Novartis called this "an important step toward personalized medicine."

Improved Research Methods Personalized medicine could improve drug development by guiding more precise trials using a few hundred patients instead of 3,000 to 5,000. "This is going to change drug development," said the CEO of a pharmaceutical company.

The complexity is staggering. Minute genetic differences must be identified and related causally to specific outcomes from a wide range of drugs. And sensitive social issues are involved. People of different social classes and races react different genetically, which is likely to provoke issues involving discrimination.

But the progress noted above is compelling. Some experts think it will be common for patients to be genetically tested at about 2010. TechCast estimates treatments tailored to individual genetic differences could enter the mainstream about 2018, saving hundreds of billions of dollars and greatly improving healthcare.

Genetic Therapy

Genetic therapy may represent the Holy Grail of medicine because so much illness is inherited from approximately 5,000 genetic disorders that have been identified. The decoding of the human genome has not reached the point where genetic blueprints have been mapped for all these diseases, and the techniques for altering genetic traits remain crude. But experiments have shown promise, as well as dangers.

A famous French trial completely rebuilt the immune systems of children, although three of them subsequently developed cancer because the virus carrying the corrective gene affected other genes inadvertently. Recently, the U.S. National Cancer Institute successfully gave two men new immune systems to combat cancer, although others in the trial died. The researcher said "I consider this proof that it can work."

More than 300 companies are developing genetic therapies involving 500 clinical trials. The journal *Nature* announced the development of "zinc fingers," an amino acid with protuberances resembling fingers that bind to defective strands of DNA and correct the code. "It deletes the miscoded DNA and fixes the problem," said David Baltimore, co-author of the paper. A harmless virus has been shown to cross the notorious "blood barrier" by carrying genes through the bloodstream into almost every muscle cell. Genes are being injected into the heart to stimulate the formation of new blood vessels that alleviate cardiovascular problems. Research is underway to treat blood disorders, Alzheimer's, illnesses of the eye, pancreatic cancer, and skin cancer.

The challenge of identifying genetic defects, delivering corrective DNA into the body, and having the new genes accepted is enormous. But people badly want to be freed of horrible genetic illnesses, and they are using influence to accelerate progress. For instance, the death of President Reagan exerted pressure to cure Alzheimer's.

Nobel Laureate Walter Gilbert thinks the genetic causes of 2,000 to 5,000 hereditary diseases will be understood by 2010, and that cures for most of these illnesses will be commonly available by 2020 to 2030. The Chairman of Amgen estimates "Gene therapy will be in common use by 2025." This agrees with TechCast's forecast that 30% of genetic illnesses are likely to be cured by 2024 +/– 5 years.

Life Extension

Opinions on aging are notoriously controversial, but the evidence seems to be accumulating that life extension is possible. Discoveries are being made in extending the life of cells, repairing damage to the body, replacing organs, curing major illnesses, and improving lifestyles. As a result, trends suggest that human life spans could regularly approach what seems to be the natural limit of 120 years. The challenges and social consequences are enormous, but many authorities are confident the problems can be solved. Let's look at some research underway:

Trends in Life Extension

Telemeres Researchers have identified an enzyme, "telomerase," that causes human cells to replicate hundreds of times beyond what was thought to be the limit of cell reproduction, the "Hayflick Limit."[4]

Resveratrol A substance in red wine – resveratrol – has been found to protect animals from illness and aging. "It's the first example of a drug that controls the aging process. Before, this would be considered snake oil," said a scientist at NIH.

Sirtuins Scientists have found that "sirtuins" are "universal regulators of aging in virtually all living organisms." These genes seems to explain the success of calorie restriction diets, which prevent age-related diseases and prolong life; now the hope is to control sirtuin genes directly to gain these benefits without dieting. Sirtris Pharmaceutials is testing a drug using sirtuins that is a thousand times more potent than resveratrol, and results for extending life spans to 100 years or so look promising. David Sinclair, the lead scientist, says "This will impact humans within a decade."

Anti-Aging Genes A study at Southwestern Medical Center identified the gene in mice that causes the negative effects of age. The Institute that funded the study said, "This could promote healthy aging and longevity in people."

Nanotech Some scientists believe that nanotechnology will provide the means to keep the human body healthy endlessly. Ray Kurzweil foresees "Fleets of computer controlled molecular tools, smaller than a cell, removing obstructions in arteries, killing cancer cells, and otherwise repairing the body."

Research Funding John Sperling, the billionaire who founded the University of Phoenix, has started an endowed research project to study life extension, "A Manhattan Project whose aim is to [solve the problem] of aging."

A particularly prominent advocate of life extension is Aubrey De Grey at the University of Cambridge. De Grey claims all causes of aging are potentially solvable and that life spans will reach 130 years by 2030. But in scathing articles, respected scientists doubt that the seven solutions proposed by De Grey are feasible. Prof. S. Jay Olshansky at the University of Chicago expects mean life spans to top out at 85 years for genetic reasons, and Prof. Leonard Hayflick at the University of California thinks, "Superlongevity is simply not possible." But studies show that such claims have consistently been proven wrong.

Nobody thinks life extension will be a panacea. Accidents, illnesses, etc., will always shorten lives, and many people say they prefer a natural life span. The social consequences of an aging population are uncertain, but they would be huge. Some studies suggest that the extension of healthy lives creates wealth.[5]

But if the visionaries are right, we could approach what appears to be the 120-year natural age limit, and possibly live much longer. One think tank estimates that a child born today will have a 40% chance of living to 150 years. The Director of the Laboratory on Longevity at the Max Planck Institute thinks life spans will increase by two to three years per decade, approaching 130 years by 2050. Integrating all this evidence, our experts estimate that average life spans are most likely to reach 100 years about 2030.

A HARBINGER OF THE AGE OF CONSCIOUSNESS

As the bubble chart shows, telemedicine is likely to enter mainstream use in a decade or so and will help alleviate the escalating costs of healthcare, followed soon by advances in artificial organs. These technologies are well-developed but await social acceptance and over-

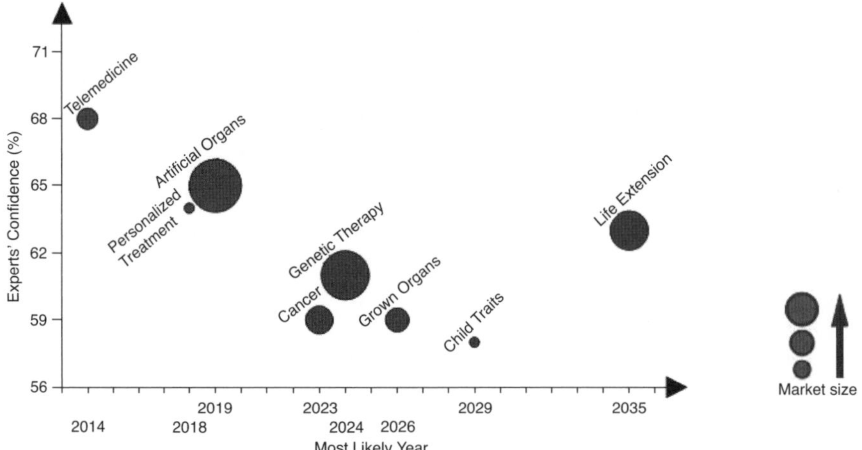

Figure 5.1 Bubble Chart of Medicine and Biogenetics

coming economic and political barriers. The need for artificial limbs, for instance, hardly constitutes a mass market, so the military is supporting research. Healthcare is likely to be more efficient and convenient fairly soon, but patience will be needed as we await the arrival of DNA-based treatments.

These DNA related technologies – personal treatment, genetic therapy, cancer cures, child traits, grown organs, life extension – all require major breakthroughs in biogenetic engineering. Apart from simple tests for genetic disposition to a few illnesses and selecting the gender of a child, very little of this can be done today. It is likely to require another 20 years or so to reach the point where biotech enters mainstream use.

A major obstacle will be the enormous difficulty of reaching accommodations on the delicate social and moral dilemmas involved. For example, the cost of sophisticated treatments is primarily responsible for the uncontrolled rise of healthcare even now, so who is going to pay for the really difficult technologies? Some argue that private healthcare systems amount to a *de facto* form of rationing used to solve this problem. What will happen when the stakes get even higher?

The problem is especially severe because relatively few people take the demanding steps needed to maintain good health. Witness the fact that one-third of Americans are medically classified as obese. I've been involved in a variety of physical disciplines – jogging, yoga, tai-chi, meditation – and I've learned that the body is a beautifully designed product of evolution but it requires a minimal level of care to function well. Many people find it hard to moderate their diet, exercise, control stress, and accept responsibility for their well-being.

And the impasse over abortion, euthanasia, stem cell research, and cloning is simply a foretaste of the far less palatable choices yet to come. Who will decide which traits parents are allowed to select for their offspring? If it were possible to genetically engineer a genius-level IQ, should parents be permitted to create this advantage for their children? How will employers, insurance companies, and potential mates welcome the news that you are genetically predisposed to a serious disease? And with research producing part-human, part-animal "chimeras" that can be used as living factories to produce organs for transplant into humans, the moral issues surrounding bioengineering become very murky indeed.

These worries pale in comparison to more troublesome possibilities. Were you shocked when cruising the information superhighway became stalled by viruses, spam, hackers, and identity theft? Wait until you see their biological equivalent – biohackers – bright kids and terrorists inventing all manner of clever biological agents that can do harm through insidious biological pathways. Were you frightened by the threat of the Ebola? AIDS? SARS? Our waste products contain so much biologically active material that the microbes seem to be evolving more quickly, requiring massive research efforts just to stay ahead of their destructive power.

The range of mischief we are moving toward is limitless precisely because the power of these new technologies is so vast. Yes, it's wonderful to gain mastery over the forces of life, to design almost any type of organism, cure almost any illness, live for hundreds of years, and so forth. And it makes sense from an historic point of view. We have now mastered almost all aspects of the physical world with atomic power, space flight, etc. Why shouldn't mastery of the living world be next?

Biogenetic engineering is likely to confer mastery of life processes, but this God-like power will require more sensitive moral awareness and intense self-discipline to manage such terrible dilemmas. In that sense, advances in medicine represent a harbinger of what is to come generally throughout the social order – the unavoidable need to cultivate a higher-level of consciousness, as we will see in Chapter 9.

Notes

1 Center for Disease Control. www.cdc.gov.
2 Robert Charette, "Dying for Data," *IEEE Spectrum* (October 2006) pp. 22–27.
3 Buchanan, *From Chance to Choice* (Cambridge University Press, 2000).
4 Stephen S. Hall, *Merchants of Immortality* (NY: Houghton-Mifflin, 2003).
5 "The Health and Wealth of Nations," *Science*. (2000) Vol. 287, pp. 1207–1209.

CHAPTER 6

Faster and Farther: Building the Global Transportation System

A great example of transportation's importance can be seen in the fabled Silk Road that once connected Europe to the Far East. During most of the first millennium, caravans made the grueling trek from Rome through Turkey and Persia, across the long, barren deserts and rugged landscape of Afghanistan and Pakistan, to reach the Tang Dynasty in Eastern China.

The Silk Road was the first intercontinental superhighway. It allowed Europeans to exchange precious metals for silk, spices, and other Chinese luxuries, and it transferred knowledge between these disparate cultures to spur development throughout the known world. Today, a new Silk Road is forming as the West is being connected to the East through jet travel, global markets, and electronic media.

THE PARALLEL DIMENSIONS OF TRAVEL AND COMMUNICATIONS

As the summary below suggests, breakthroughs in car design should take us faster and further on a new fleet of "lean and clean" hybrid and fuel cell cars, guided by intelligent systems zipping along automated highways. Small aircraft are likely to become a serious alternative to commercial airlines, maglev trains floating on air at 400 mph may replace short flights, while hypersonic flights will carry us across the globe as easily as we now cross a nation. It's going to take two decades or so, but this global transportation system is emerging now to serve the needs of nine billion people living on a planet that is being integrated economically.

Summary of Transportation Forecasts

Hybrid Cars This first alternative to the internal combustion engine is revolutionizing auto and truck designs and is likely to become a major force in the global car market.

Fuel Cell Cars Hydrogen fuel cells seem destined to become the next generation of car engines beyond the hybrid, powered by alternative energy sources.
Intelligent Cars Global Positioning System (GPS), navigation, traffic monitoring, collision avoidance, and automatic functions are creating intelligent vehicles that improve safety and convenience.
Automated Highways Increasing traffic congestion is likely to be ameliorated by caravanning platoons of cars along electronically controlled highways at 80 mph.
Small Aircraft Small planes and helicopters are becoming cheaper and easier to fly, offering enormous convenience for millions of people.
Maglev Trains Propelled by magnetic coils and riding on a cushion of air, maglev trains are beginning to connect major cities at 300 to 400 mph.
Hypersonic Planes Beyond the Concorde, a new breed of high-performance airliners flying at Mach 10 should reduce travel across the globe from 30 hours to 3 hours.

But why is all this travel needed, given the dazzling information systems described earlier? With modern homes and offices becoming integrated through the glories of high-bandwidth, perfect fidelity, interactive audio and video, you would think we could replace a lot of travel with really good communications. One of our experts, Dennis Bushnell, Chief Scientist at NASA Langley, states the case well: "Immersive presence, virtual reality, holographic projections, and other forms of "virtual travel" will win out over physical presence for many reasons."

This may be true, but we are also learning that the physical and virtual worlds co-exist in parallel dimensions. There seems to be a constant need for travel to exchange goods, to experience the feel and function of merchandise, and to interact with others on a fully human scale. The most brilliant, most powerful IT system may never replace this primary need. That's why the amount of time spent traveling has remained fairly constant throughout history. As seen in Chapter 4, far more sophisticated IT will certainly bring more vivid online experiences, but it will remain a convenient although limited alternative to the real thing.

Hybrid Cars

Under the leadership of Japanese carmakers, hybrids have become a successful alternative to the internal combustion engine. Powered by a small gasoline engine operating only at high speeds where it is most efficient and a battery at lower speeds, they provide as much as 70 mph fuel efficiency and 90% reductions in pollution. Diesel engines work even better, and fuel cell versions may be coming soon, as noted below. Equipped with high-capacity batteries, "plug-in hybrids" can recharge at night with cheaper energy, raising performance to the equivalent of 100 mpg. All these designs use electric motors to drive wheels, the battery to store energy, regenerative braking to conserve it, and may in time use composite bodies that are lighter. Volkswagen built a hybrid using light-weight carbon fiber composite metal, increasing fuel efficiency to 239 mpg.

Hybrids were uneconomical at first, costing $3,000 more than conventional cars, while the saving in fuel is only about $300/year. This caused an auto executive to say "The majority of Americans don't give a damn about hybrids."

Sales are increasing dramatically, however, and the added cost is dropping. Toyota's upgraded second Prius is comparable to a Camry in size, price, and performance, and demand is so great that some buyers are paying premiums of several thousand dollars over the list price to avoid long wait times. Toyota is building hybrid versions of its entire product line at costs that match ordinary cars, Ford is developing hybrids for its product line, and GM is building one million hybrids. Because large vehicles use half of all gas in the U.S., the big target is likely to be SUVs, buses, and trucks.

Cars last roughly 15 years, so it takes a long time to change the fleet of vehicles. Hyrbids now make up about 2% of new car sales globally and J.D. Powers expects sales to reach 4% by 2010. William Ford, Chairman of Ford, thinks hybrids could make up 75% of the car market by 2025. If present indications hold, our experts estimate 30% adoption in new car purchases about 2013.

Fuel Cell Cars

Fuel cell autos seem to be the logical follow-up to hybrids. Because of their great advantages, the auto industry is pouring billions into research, developing prototypes, and reducing costs. Internal combustion engines are only 14% efficient, while hydrogen fuel cells are 42% efficient. Finding a cheap, energy efficient supply of hydrogen is problematic, but methods are being developed that could prove feasible fairly soon. (Also see Alternative Energy)

The challenge is considerable. Some estimate the cost of the fuel-cell system alone at $30,000, and the cost of hydrogen is presently several times that of gas. Methods for extracting hydrogen are still being developed and are not yet commercially feasible. For instance, clean water must be used for hydrolysis because the energy required to split sea water into hydrogen increases ten-fold. Some scientists warn that hydrogen from fuel cells could leak into the atmosphere and disrupt the ozone layer, and others claim producing hydrogen from oil and coal would release more carbon dioxide. Gasoline refineries, gas stations, distribution systems, and the entire infrastructure of auto fuel would have to change.

But the Fuel Cell Council claims use of fuel cells is growing 75% per year as these problems are addressed. Here are some key developments:

Fuel-Cell Car Developments

Toyota and Honda rolled out the first fuel cell cars in 2002 at a cost of $1 million each. With costs falling to $100,000 now, Japan is planning to have 50,000 fuel cell cars by 2010 and five million by 2020.

Germany leads the world in this field and was first to produce hydrogen using electrolysis. Daimler is introducing a fuel cell car in 2010 that should match the cost of ordinary cars.

GM has designed a special frame for fuel cell autos and plans to have 40 cars out soon. The company is investing $1 billion/year in hydrogen research and intends to become the first to sell one million fuel cell cars. A GM official said "Fuel cells are the first technology in 100 years to rival the internal combustion engine."

Ballard, the world's leading maker of fuel cells, is planning to reduce costs from $2,000/kW to $50/kW by mass producing 300,000 fuel cells.

Munich opened the first automated liquid hydrogen fueling station in 1999, and other stations are in use now.

The California Fuel Cell Partnership of car makers and energy firms plans to put 300 fuel cell cars in tests and build fueling stations on major highways.

Government Support is strong. The U.S. is putting $1.7 billion into hydrogen research, the E.U. is investing $2 billion, and Japan $2 billion.

Iceland is building the world's first hydrogen economy using its abundant geothermal energy to create hydrogen.

China is hungry for energy, and may take the lead by committing to fuel cell development.

Hydrogen is merely a carrier of energy, so the big problem is developing economic and environmentally safe methods for creating and storing hydrogen. Roughly 250 fuel cell systems were under development in 2006:

Methods for Producing Hydrogen

GE has a working prototype of an electrolyzer that splits water to form hydrogen at prices comparable to gasoline.

Pacific Northwest National Laboratory is developing an under-the-hood reformer that produces hydrogen from gasoline at high efficiency.

Cientifica, a leading nanotech company, uses carbon nanotubes to build fuel cells with a ten-fold improvement in performance and cost.

Millennium Cell Company uses sodium borohydride to produce hydrogen on demand. The technology is expected to be ready about 2008.

Britain Wind turbines are being used to split water into hydrogen directly at competitive rates. "We can produce hydrogen for about the same price as petrol," said a British researcher. Two British firms use solar cells that excite water to split it into hydrogen, reducing costs by a factor of four to ten.

Honda has built an entire fueling station that uses photovoltaic solar cells to split water, the model of green energy production.

Penn State University uses special bacteria to coax four times as much hydrogen from biomass than by fermentation alone.

The Weizmann Institute in Israel developed a system that uses Magnesium or Aluminum to split hydrogen out of water on board a car at low cost.

Oxford University scientists replaced the metal catalyst used in most fuel cells with enzymes from bacteria and fungus to make fuel cells that are smaller, cheaper, and simpler.

University of North Carolina researchers have developed a proton-exchange membrane that increases electric flow in a fuel cell two to three times more than current membranes. Plans are to improve this by another 20 to 40 times.

Prof. Daniel Sperling, an authority on fuel cell autos, forecasts: "Conventional wisdom is that fuel cell vehicles will progress by 2010 to where hybrids are today, selling hundreds of thousands/year." TechCast estimates that fuel cell cars will enter the commercial market about 2013, roughly the time hybrids reach mainstream use, passing the torch to the next generation of auto designs. In time, global demand could reach several trillion dollars.

Intelligent Cars

Traffic is not going to diminish in the Information Age, but IT is being used to create "intelligent cars" that guide drivers through traffic congestion, improve safety, serve as a communications hub, and perform a variety of automatic functions. As noted under "Automated Highways" below, the wasted fuel, pollution, lost work hours, and accidents caused by congestion cost the U.S. more than $200 billion per year. Millions of drivers have used intelligent auto systems in Europe and Japan for years. The potential for transforming the auto industry is so attractive that the government of Taiwan has targeted this as the next trillion dollar market to develop.

The intelligent car poses an interesting chicken-egg problem because it makes little sense without intelligent highways and *vice versa*. To complicate matters further, common standards must be developed and accepted so that cars are compatible with infrastructures. Advances are coming in both spheres.

In addition to mobile communications, GPS, cruise control, collision-avoidance, and other common features, Volkswagen has built the world's first fully automatic car. The firm's new Golf uses radar and lasers to monitor the road, GPS to pinpoint the car's location to within an inch, and automatic controls to steer, brake, and accelerate. Nissan has developed a "Lane Departure Prevention" system using video cameras and computerized controls that buzz when the driver leaves the lane and then nudge the wheels to correct. Toyota has developed an "Intelligent Parking System" that uses video cameras to park the car automatically. GM and other car makers are introducing "fly-by-wire" steering that replaces the mechanical steering wheel with a computerized joystick. All GM cars will have OnStar installed soon.

Comparable progress is underway on the infrastructure. Japan has launched the world's first national intelligent highway system, and Ontario is building one also. The U.S. government is installing traffic monitoring systems on 20,000 miles of highway in 108 of the largest metro areas, while companies like Traffic.com and Inrix provide traffic forecasts for the largest U.S. cities. And intelligent traffic lights are being developed that sense heavy traffic and adjust the light sequence to optimize timing. All this, combined with electronic toll collection and other systems, are forming the basis of an emerging infrastructure that will increase road capacity 30% and reduce commute times.

With the problem being attacked from both the auto and the infrastructure ends, costs are dropping and usage is rising. The average price of a complete intelligent car system is expected to decline from $2000 to about $1000. In 2005 only two million out of 18 million new cars sold in the U.S. had intelligent features, but that should increase fivefold by 2010. TechCast's best estimate is that intelligent cars are likely to enter the mainstream (30% adoption) about 2010 to 2015.

Automated Highways

With no end in sight to mounting traffic congestion, automated highways offer a far less costly, faster, and safer way to travel than building more roads. Automated highways would increase capacity, and they cost less than $10,000 per mile, compared to $1 to $100 million per mile to build new highways. Cars would be equipped with sensors and wireless communications to control speed, steering, and braking on electronically equipped lanes, spaced at close intervals to form convoys or platoons. General Motors has tested cars traveling at 70 mph bumper to bumper with no drivers.

True, the thought of careening down the highway in close formation with scores of other cars does not inspire confidence. But after a bit of reflection, most people conclude they would rather depend on the computer systems than other drivers. It's also true that some drivers are reluctant to yield control over their driving, but many others are delighted to be freed for reading and doing other things while traveling.

There are other barriers. Technical feasibility is limited to a few controlled test track demonstrations. And who would be liable for accidents on automated highways? A BMW executive thinks, "The vision that a machine can replace the driver is far, far away in the future. No technical system available today is as reliable and as flexible as the human being."

But the rise in traffic makes automated highways almost inevitable. Average speeds on commuter corridors in the U.S. are dropping as traffic increases, while congestion costs $74 billion per year and is expected to double by 2020. U.S. highway crashes cause 40,000 fatalities and five million injuries each year, costing $150 billion. Traffic engineers estimate automation would double or even triple highway capacity, improve safety considerably, and reduce highway construction costs. Because of steady, controlled speeds, automated highways could even reduce fuel consumption and exhaust emissions.

The Japanese and Europeans are working with carmakers on various automated highway systems. Japan has even introduced a system that automatically returns rented cars, and smart taxis without drivers will hit the market soon. In the U.S., an initiative is being launched that uses automatic systems to reduce collisions, and carmakers are adding radar, cameras, and adaptive cruise control systems that maintain spacing, as noted in "Intelligent Cars" above. Terry Duncan of Ford says, "All the technology is in place right now. Older people today may not trust their car to steer for them, but I don't think young people would have a problem."

The U.S. Federal Highway Administration estimates that intelligently controlled highway systems are likely to be available about 2010 to 2020. TechCast is a little more cautious and forecasts their use on 30% of major highways by 2025 +/– 5 years.

Small Aircraft

Although we have long considered the idea that people would fly to work in private aircraft a bit fanciful, this outrageous idea is now a reality for many, and it could sweep through modern society. Two trends make it possible. Costs of owning small aircraft are falling dramatically, and technology now makes piloting a small plane almost as easy as driving a car. That's why sales of helicopters and personal jets are soaring in many countries, and trends suggest that private planes may increase to the point where masses of airborne commuters rival commercial travel.[1]

The attraction, of course, is convenience. Commercial air travel is congested and fraught with delays, and the U.S. Federal Aviation Administration (FAA) expects air traffic to double by 2010, making private flights attractive. A typical one hour commute can be reduced to a few minutes, often flying directly from home to office. A small business owner in Minneapolis parks his chopper in his home's backyard and his company's parking lot. That's convenience.

Cost has dropped to the point that a good plane can be bought for $150,000 – not much more than an expensive auto. Operating costs have also dropped such that a short trip only costs roughly $10 – again, about the same as driving your car. And economic growth has created a large number of affluent professionals who find the convenience well worth the cost. A Seattle businessman commuted to work on a friend's aircraft, reducing a 60 minute drive to 6 minutes. "I immediately bought one," he enthused. "You don't have to be a millionaire to do this."

Safety is also improving. Accidents plunged 90% since the shift from prop engines to jets, and some new planes have their own parachutes that can float the craft down gently if necessary. Because today's small planes are easier to control and are equipped with GPS, average people can become trained to fly in a few months. Inventors are designing airplanes with creatively simple features rather than the conventional wing-body-tail, cutting costs and improving stability. One plane is powered by electric motors and hydrogen fuel cells, and another version uses solar cells.

The infrastructure is also developing as airports appear at strategic locations. Wall Street now has a heliport, as does Philadelphia. In the U.S., 98% of Americans live within 30 minutes of 5,400 small airports that handle 37 million flights annually.

To handle this onslaught of small aircraft, the FAA has been experimenting with "free flight" rules for years. Aircraft are equipped with GPS, radar, and collision avoidance systems, so they can monitor their location *vis-à-vis* other aircraft and maintain safe distances. The FAA finds that free flight is safer than controlled flight and it can handle vastly increased air traffic. NASA is planning a "Highway in the Sky" system in which computerized navigation controls are displayed "like a video game."

With these supporting trends at work, small aircraft are entering the market in droves. Sao Paulo, Brazil, is home to more than 500 private helicopters used to take executives to work. The Robinson R22 is the world's most popular helicopter, with thousands sold every year. Honda is developing an air taxi designed to ferry four to five people between airports. Eclipse Aviation has 2,100 orders for its two engine, five passenger "microjet" costing $1 million, one-fourth the cost of a corporate jet. Cessna and Adam Aircraft are building similar "personal jets." Moller International is developing a "Skycar" that travels at 350 mph and gets 15 mpg fuel efficiency.

A note of caution is called for however. The prospect of endless small aircraft flooding the skies invites disaster, as seen in countless crashes that often destroy buildings and kill many people. How can thousands of small aircraft be controlled in metropolitan areas? What would happen in the event of breakdowns, and are we prepared for the rise in death tolls? Would we see fuel, broken parts, and other junk rain from the sky occasionally?

Entrepreneurs think 20% of business travel could move to 35,000 small "air taxis" in 2010, and NASA estimates a market for 8,300 microjets by 2010. "It's going to be a revolution in the transportation industry," said a NASA spokesman. TechCast thinks small aircraft are likely to be used for 30% of air travel by about 2025 +/– 5 years.

Maglev Trains

Maglev trains use electric coils to generate magnetic fields in the train and the track to lift the car above its rails and propel it forward at high speeds. The concept is attractive because traveling at 350 mph or more is a convenient alternative to short flights between large cities, and trains provide immediate access to city centers. Maglevs have been tested and debated for decades, but China's introduction of the world's first commercial train may determine the future of this technology. Critics claim there is little advantage over present high-speed trains and the costs are too high, while others point to the huge gain in comfort, prospects for still higher speeds, improved efficiency, less pollution, and cheaper operating costs.

Costs are estimated at $50 to $75 million per mile, and conventional trains are reaching similar speeds, which diminishes the advantage of maglev trains. The French high-speed train reached 300 mph, almost matching the speed of maglev trains. One critic claimed maglev "could turn out to be the Concorde of railways." That's why Britain is replacing its maglev plans at Birmingham Airport with shuttle buses. Japan invested $2 billion in a maglev to connect Tokyo and Osaka but is reluctant to complete the project because bullet trains provide three-hour service. Germany canceled a $6 billion plan to connect Berlin and Hamburg.

Some experts claim maglev does not have to be more costly, that operating costs are dropping, and there are great advantages. Maglev trains are less polluting, they use one-third less energy than conventional rail and 70% less than high-speed trains and airplanes, and they are more comfortable because they ride on a cushion of air. One Japanese passenger said "I would use the maglev rather than an airplane."

In 2003 China launched the world's first commercial maglev running between downtown Shanghai and the city's airport. The train is expected to carry 20 million passengers by 2010. Another $15 billion project is being planned to connect Shanghai and Beijing, reducing travel time from 17 hours to 4 hours. Los Angeles (LA) is funding a maglev project connecting the city to airports. "This is technology for the 21st century," said an LA politician. "The time has come to start using it."

The U.S. is spending $6 billion to build a maglev system between Washington, DC, and Baltimore, and another in Pittsburg. The Secretary of Transportation said, "These two projects demonstrate what may become the backbone of transportation in our densely populated corridors."

The technology is improving, and the fruits of China's effort should become visible soon, which could spur investment. An improved version of maglev, called "inductrack" uses passive magnets rather than electrically generated fields; it is less expensive to build, maintain, and operate. Japan set a new speed record of 360 mph on a maglev train recently. TechCast estimates 30% of heavy traffic corridors in industrialized nations are likely to use maglev trains by 2030 +/– 10 years.

Hypersonic Planes

The French-British Concorde may be passé, but the prospect of hypersonic flight is alive and well as globalization drives increased travel across the world. A new generation of hypersonic planes is under development in the U.S., Russia, Australia, and Italy that incorporates lightweight airframes made of exotic metals and reliable scramjet engines, which suck in oxygen from the thin atmosphere at high altitudes. Materials more heat-resistant than insulation tiles have been developed to keep hypersonic aircraft from burning up, and computerized controls are being used to solve daunting flight-dynamics problems.

Russia is testing hydrogen ramjet engines flying at Mach 8.5. Aerospatiale Matra is developing a new supersonic transport that could be flying by 2015, followed by a 1,000-seat flying wing in 2020. NASA is testing a scramjet aircraft capable of Mach 7 to 10 at 95,000 feet for advanced space missions, while UTC Corp. and Purdue University are developing propulsion systems for Mach 7 flight. The "Supersonic Aerospace International" is being built to fly at 1,100 mph, use normal runways, and avoid sonic booms. It is expected to enter service in 2012.

As governments and corporations gear up to deliver these vehicles over the next decade or two, it should be possible to reduce flying time from the Eastern U.S. to Asia from 30 hours to 3 hours. TechCast experts think hypersonic planes are likely to be used for 30% of long flights by about 2028.

NO REST FOR THE WEARY ROAD WARRIOR

As the bubble chart shows, we can expect to see hybrid cars enter the mainstream in a few years, with the first fuel cell cars being sold

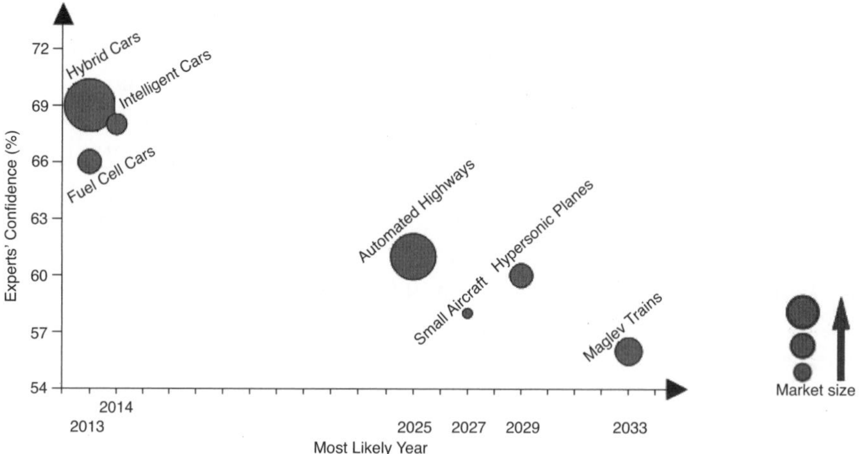

Figure 6.1 Bubble Chart of Transportation Forecasts

commercially soon after. Intelligent systems will likely proliferate to make autos more automatic, convenient, and safe, leading to automatic highways that speed traffic along heavily traveled routes in two decades or so. Small aircraft and hypersonic commercial carriers are likely to be widely used about this same time, and maglev trains are likely to connect major cities.

All in all, it should prove a sophisticated new transportation system for a global economy. Our earlier forecasts in Chapters 2 and 3 showed that globalization will likely reach a critical point about this same time period of 2020 to 2030 as India, China, and other developing nations become modernized. With as many as five billion participating in a global economy, we should see roughly a five-fold increase in transportation needs, a large part of which will be intercontinental. The breakthroughs noted above in hybrid and fuel cell cars will be driven by the need to avoid commensurate increases in environmental loads and energy needs. Small aircraft, intelligent cars riding on automatic highways, hypersonic flights, and maglev trains will prove essential to carry this burgeoning mass across nations and continents.

While this should be a more robust transportation system that permits further travel at faster speeds in greater convenience, the basics may change very little in a human sense. Most people are likely to expend roughly the same amount of time and hard work traveling for work or leisure. They will simply visit more distant places more easily. Before industrialization, people in agrarian societies walked and rode horses within a 20 to 30 mile radius of their birthplace. Today, we think nothing of driving across entire states and flying to nearby nations. The coming generation of transportation systems will simply move our travel habits up a notch to span the globe. Travel may prove more exciting and adventurous, but it will also be more demanding. Sorry, but the future offers no rest for the weary road warrior.

Note

1 James Fallows, "Free Flight," *Public Affairs* (2001).

CHAPTER 7

The Final Frontier: Preludes to Deep Space Travel

Space is the next great frontier in the rise of civilization. The invention of primitive tools created the early world of humanity, agriculture permitted the rise of civilization, mass production produced material wealth, and now information technology (IT) and knowledge are leading to a more powerful domain of consciousness, as we will see in a following chapter. Our forecasts suggest that a number of path-breaking space projects should open up the limitless expanse of the universe over the next few decades – though we have only the faintest glimmering of what lies beyond our solar system and what our entry into deep space will mean.

EXPLORING THE SOLAR SYSTEM AND BEYOND

Several long-term trends are moving science and technology into this new frontier. One of the most powerful forces is the imminent privatization of space. We are accustomed to thinking only large governments like the U.S., the E.U., and Russia have the wherewithal to finance and manage these huge undertakings, but private space ventures are poised to eclipse the public sector. Even now the majority of satellites are launched into earth orbit by private companies, and tourist flights are planned that promise to make commercial space travel a reality within a few years, as we shall see below. As the thrill of experiencing space becomes available to almost anyone, it is easy to imagine the floodgates of free enterprise opening wide to permit a rush of exciting space adventures – roughly the way the West was tamed by hordes of hungry pioneers a century ago in the U.S.

Another powerful trend is the relentless improvement in space technology, much the way IT has been skyrocketing. This chapter will show that many of the technologies supporting space travel – microelectronics, sensors, control systems, exotic materials, telecommunications, etc. – are accelerating in performance at roughly the same astonishing rate as that of computers, DNA sequencing, and many other fields, doubling in performance every two years or so. Propulsion systems able to travel faster are also advancing, but not at the same exponential rate.

And the "virtuous cycle of knowledge" described in Chapter 1 is revealing a grand new scientific perspective of the universe, making space travel the next logical step in technological progress. Science is learning how stars are born, evolve, and die; how planets are formed; the composition of comets and other celestial bodies; the terrain of the Moon, Mars, and other planets; how to manage planetary orbits and climate, and most other features governing the universe.[1] We are becoming so familiar with space that the inventor of the most popular computer game – *SimCity* – will soon be releasing *Spore* – a game that allows you to create entire galaxies and planets, seed them with life, and guide civilization's growth. It should feel like playing God.[2]

These forces should push space technology toward ever greater reach, possibly leading to massive projects that dominate the attention of modern societies in about 30 to 50 years – a true "Space Age." Here's a quick summary of the breakthroughs that should be most prominent. As always, remember that detailed references are available at www.TechCast.org.

Summary of Space Forecasts

Space Tourism With spaceports being built and NASA's blessing secured, it's highly likely that private spaceships will carry tourists around the Earth in a few short years.

Moon Base NASA is working now to establish a permanent colony on the Moon, in preparation for a manned trip to Mars. It should prove to be a landmark event.

Men on Mars Despite the arduous challenges involved, we think a manned landing on Mars is likely shortly after the moon base is operational, about 2028.

Contact There is huge unpredictability in contacting alien life, but the technology is advancing very rapidly, and scientists estimate the odds are overwhelmingly favorable.

Star Travel Some experts are doubtful, and major breakthroughs in physics are needed, yet we think humanity could be launched into deep space about mid-century.

Space Tourism

People thought the idea of "space tourism" a bit silly at first, but now that a few wealthy souls have traveled into orbit, it is suddenly almost certain that many more will follow. The lure of private space flights was dramatized when Burt Rutan's SpaceShip One made its second flight into space to win the $10 million Ansari Prize. Business people think the huge costs could be reduced by a factor of 10 or more, and that thousands if not millions are prepared to pay for the privilege of going into space. One entrepreneur said, "Business is looking hard at space tourism; they don't want to miss an opportunity." Peter Diamandis, who managed the Ansari Prize, thinks, "We're on the verge of a golden age. It should soon be possible to tour space on a routine basis."

A number of market surveys show people are interested. Zogby conducted interviews of 450 wealthy people and found that 19% would likely take a suborbital flight 50 miles into space at $100,000. Seven percent were willing to pay $20 million for a two-week orbital flight to a space station. Other market research shows 10,000 people would pay $1 million or more for the experience of traveling to space. A study by Futron estimates that 15,000 suborbital flights and 60 orbital flights will take place annually by 2020.

With the technology readily available and paying customers waiting, it's little wonder that business is rushing to accommodate them. The U.S. Federal Aviation Administration (FAA) has approved proposals to build private launch facilities in New Mexico, Oklahoma, and Texas. The New Mexico facility has signed a $225 million contract to support Virgin Galactic, Richard Branson's new venture, which has thousands of clients eager to take suborbital flights for $200,000. Burt Rutan has partnered with Branson by building SpaceShip Two for the suborbital flights Virgin Galactic is offering. Rutan says the flights will be as safe as airlines.[3]

Space Adventure Company is planning to take tourists to the International Space Station for $20 million and around the moon for $100 million within a few years; they claim 1,000 clients are interested in taking the flights. Billionaire Robert Bigelow is spending $500 million to build a space hotel and is planning a cruise ship ferrying tourists to the moon. His chief scientist claims, "We've got the know-how, and he's got the money," while a chief scientist at NASA says, "We need more Bigelows."

NASA thinks tourism is expected to begin about 2010 to 2012, and that a space hotel might be built in two decades. The FAA has already licensed the first commercial space flight. TechCast thinks we should see the first "space cruiser" take a group of tourists on an orbital flight around the Earth sometime around 2014 – a mere seven years from this writing. A suborbital flight could be launched sooner, possibly about 2010. Less than 10% of our experts think this will never happen, which implies more than 90% of our best minds collectively estimate that ordinary people will travel into space on private flights in a few years.

Who would have believed it a mere five to ten years ago? Most importantly, what will humanity learn to do next?

Moon Base

The Moon has long been considered a convenient launching pad for space missions. It is relatively close, easy to escape its mild gravity, and has almost all materials needed for other space ventures, including water. This logic prompted President George W. Bush to announce a $12 billion plan to establish a permanent settlement on the moon, leading to a manned mission to Mars.

Since then, NASA has set up an Office of Exploration Systems to develop the Constellation system for a Moon mission. By 2008, the crew exploration vehicle (CEV) that ferries astronauts to and from the Moon will be tested for the first manned mission in 2014. New rocket motors, hybrid fuels, and other technologies are being developed to create more powerful launch vehicles. TransHab, a multi-story inflatable habitat, is being designed to provide semi-permanent housing for up to six astronauts. The two poles of the Moon are being considered as prime locations because they are constantly in sunlight, allowing photovoltaic cells to collect energy. The E.U. has its own plan, Aurora, to construct a permanent Moon base from which to launch Mars missions in collaboration with the U.S.

The challenges of building and maintaining a permanent base on the Moon are daunting, but not much more so than the International Space Station, which has been inhabited for years. Like all space projects, it is hard to justify the huge costs. The president of an aerospace company thinks, "It would be egocentric to focus on a Moon colony. It's not the right time." It's also true that long-term living conditions on the moon may be demanding, physically and psychologically.

But the advantages also are considerable. A well-functioning Moon colony could serve as a far more convenient connecting point for other ventures into space. It would make missions more feasible, facilitate their launch and control, and serve as a spaceport for the coming wave of space pioneers. Economically, a Moon base would be such a huge project that it would stimulate business development among thousands of contractors, create exciting new jobs, and provide unusual opportunities for developing medical and industrial research. The Apollo Program, which sent astronauts to the Moon in 1968, launched a wave of progress, and establishing a permanent Moon colony would be a far bigger undertaking offering far greater benefits.

Experts tend to agree the U.S. will return to the Moon about 2015 with the goal of establishing a colony. NASA plans to test its new systems by sending a team to the Moon in 2018 for up to six months. Some estimate it will require as long as 30 years to establish a permanent base. These doubts are reflected in a wide variance of +/– 12 years in the TechCast forecasts and relatively low confidence of slightly over 50%. Our best forecast is that a permanent Moon base is likely to be operating on a continuing basis sometime around 2028, and that it will be part of the more ambitious attempt to land humans on Mars. NASA's official estimate is 2022 for the Moon base, so the TechCast data suggests we may see some slippage.

Men on Mars

Mars is so far away that it would take a full year for a round trip, requiring sophisticated life support systems beyond the state-of-the-art. The rockets needed to send humans to Mars are not yet available, and precautions are needed to protect astronauts from a six month journey fraught with life-threatening risks. They would suffer loss of bone density from weightlessness, and possibly as much as 40% loss of brain cells from intense cosmic radiation belts. Some estimate that it would require surrounding the crew quarters with five feet of water to protect them from radiation. Still, the symbolic allure of sending humans to a neighboring planet is hard to resist, and a variety of programs seem to be leading in this general direction.

The 2004 explorations of Mars by unmanned rovers are preparing the way by gaining a thorough understanding of the Mars terrain. NASA is sending a series of "scouts" (planes or balloons) to explore Mars, and in 2009 they plan to send a new generation of smart landers capable of exploring hard-to-reach sites. The European Space Agency is designing the ExoMars robotic mission to be launched in 2009. Between 2009 and 2011, samples of Martian soil will be brought to Earth. NASA unveiled Project Prometheus, a $3 billion effort to develop nuclear propulsion for missions to the outer planets, and it has awarded grants to develop advanced life support systems for sustaining human colonies on Mars. Russia is conducting experiments to simulate a manned mission to Mars.

If all goes well, NASA is planning to send a six-person crew to Mars by 2020. The vehicle would be loaded with 80 tons of hydrogen that is heated by nuclear thermal rockets to generate power for the journey. There are even discussions of the possibility of "terraforming" the planet to make it habitable, like Earth.

But a Mars landing is fraught with the same uncertainty surrounding the Moon base. Some authorities think 2014 is the optimal time for the first manned mission to Mars. The European Community thinks a manned mission to Mars would occur by 2035. TechCast estimates the most likely timetable for a Mars landing is roughly 2030, a few years following the establishment of a Moon base. Less than 9% think these events will "never occur," so there is good consensus that the combined Moon-Mars missions are likely to be launched in about 20 years, although both projects have roughly the same wide variance and modest confidence.

Contact

The search for extraterrestrial intelligence (SETI) was once considered quack science, but the National Academy of Science recently endorsed it, stating that "SETI is an important national resource in astrobiology." Planets are routinely being discovered orbiting stars, evidence is appearing to support the existence of life throughout the universe, telescopes and communications are becoming more powerful at an exponential rate, and scientists say the odds strongly favor contact.

The obstacles are enormous, and the very process of contacting alien life seems random and unpredictable. Communication must occur across light years of space, penetrate alien cultures, and intersect at different points in the evolution of different civilizations. Even if we found a signal, a normal conversation seems impossible. With such great distances measured in years of light travel, it's hard to imagine a two-way exchange with many years lapsing between sending a message and receiving a reply.

Some scientists think the majority of planets are hostile to life. Roughly 250 planets have been found in other star systems as of this writing, although most are gas giants like Jupiter that are likely to be lifeless. Gas giants are easy to spot because of their size, of course, while small planets like Earth are not. But there is evidence to suggest that Earth-like planets are likely to accompany gas planets. The HARPS telescope in Chile is starting to detect Earth-like planets, and studies suggest it may be routine for gas giants to be part of solar systems that have rocky, wet planets. "Planets like ours must be more common than gas giants," said a Princeton physicist.

What's interesting is that the consensus of scientific opinion suggests the conditions favorable to life are likely to occur on one-tenth of all stars. A study concluded: "The most conservative prerequisites for life are very, very basic." The concept of "panspermia" – the theory that the seeds of life are prevalent throughout the universe – has been supported by experiments that collected samples from space and found live cells at higher altitudes, well above contamination from Earth. Other studies show that organic molecules exist in deep space.

Further, scientific estimates of the chances of contacting life (the Drake equation) are overwhelming. One scientist said: "There are ten-thousand to one-hundred-thousand advanced civilizations in our galaxy alone." Another thinks: "There could be a billion Earths in the Milky Way. A third estimates: "There are twenty billion Earth-like planets in our galaxy." One study reported that Earth is in a hospitable zone in the Milky Way and our solar system is younger than most, suggesting that contact with advanced civilizations in our galaxy may be likely.

Most promising is that the technology for contacting alien cultures is improving dramatically. The capacity to detect other planets is doubling every 18 months. After all, telescope design includes the same technologies used in microelectronic chips, computers, telecommunication systems, etc., which improve performance following Moore's law. Some examples:

Advancements in Space Telescopes

The James Webb Space Telescope, now under construction, will be 2.5 times the size of Hubble, at one-third the weight and will orbit 1.5 million kilometers above Earth.

A new form of "adaptive optics" is able to extract images of planets out of the atmospheric blur that prevented earlier observations.

NASA's Terrestrial Planet Finder – an orbiting 8-meter telescope – should be able to observe Earth-sized planets at about 2012.

The Allen Telescope Array consists of 350 telescope dishes spread over 90 acres. It is expected to enable more powerful search for extraterrestrial life by 2008.

The European Space Agency is building a flotilla of seven spacecraft orbiting precisely to form a giant mirror telescope array spread out across a diameter of 100 meters.

The first optical telescope dedicated to the search for alien life (Optical SETI) has been designed at Harvard University.

Because of the enormous difficulty and uncertainty over contact with alien civilizations, TechCast's experts offer wide-ranging forecasts and have only modest confidence in their estimates. Indeed, one-third of our experts think it will never happen. On the other end of the spectrum, Seth Shostak, the senior astronomer for the SETI Project, thinks technological advances are likely to detect intelligent life in about two decades, and physicist Freeman Dyson thinks "Evidence of extra-terrestrial life will be found before 2056" and will be "The biggest breakthrough of the next fifty years."

If the scientific opinions and technological trends noted above are reasonable, we might be surprised to see that contact could occur at almost any time. TechCast's best forecast is that intelligent life is likely to be reached about 2070, but within a range of +/– 25 years. It is hard to conceive of a more historic event. The discovery of the New World will pale in comparison.

Star Travel

Despite decades of alluring sci-fi shows like *Star Trek*, the prospect of humans traveling outside of our solar system is so bleak that it is not remotely feasible in the near future. The distances are too vast, the travel time too long, the technology too sophisticated, the risks too great, and the social needs unclear. As the world unifies into a fully developed, stable community over the next 30 to 50 years, however, attention should shift to this ultimate challenge. TechCast has been consistently forecasting for decades that deep space travel is likely to occur roughly at the mid-point of this century.

The nearest star system, Alpha Centauri, is four light-years away from Earth, so using rockets traveling at 10% the speed of light (an optimistic projection) would take 40 years one way. And that's the nearest star system. Our Milky Way galaxy is 100,000 light-years across, and the nearest galaxy beyond it is two million light-years away. These are enormous distances, even traveling at the speed of light, which science believes to be impossible.

Travel to other star systems seems feasible only if breakthroughs in physics reveal more powerful forms of travel able to span these enormous distances. The problem is further compounded by the absence of any reason to pursue such a quest. The world is going to be pre-occupied with the massive global crises outlined in earlier chapters for the next 30 years – economic development of poor nations, energy and the environment, climate change, cultural conflicts, etc. – that will demand attention.

Nevertheless, early signs of movement in this direction are visible. NASA has a small but active program devoted to exploring alternative concepts in space travel, such as worm holes. The European Space Agency (ESA) is studying hibernation states that would allow astronauts to sleep through the decades-long journey to distant star systems; recent breakthroughs have put dogs into a state of hibernation for months with no ill effects. Research on gravity fields, controlling the speed of light, the nature of the universe, and other exotic topics may provide breakthroughs that permit travel to other stars, as we will soon see.

The unrealized power of a global community may surprise us. As we will see in Chapter 10, at about 2050 the world should coalesce into a global order of roughly seven billion educated people. IT networks will integrate all this knowledge into some sort of global intelligence, which could then be directed to serious space exploration. Michael Griffin, head of NASA, said, "In the long run, a single-planet species will not survive. One day, there will be more people living off of Earth than on it." The famous physicist Stephen Hawking, said much the same: "Survival of the human race depends on finding new homes elsewhere in the universe… we could have a permanent home on the moon in twenty years and a colony on Mars in the next forty."

Our data forecast that humanity will send a mission to another star at 2065 +/– 35 years at a 49% confidence level. Twenty-five percent of TechCast experts think it will not happen. TechCast has been studying this prospect for almost 20 years and our most striking observation is that many forecasts invariably focus on the middle of this century. Based on these observations, my best judgment is to modify the TechCast data a bit by estimating that a space ship is likely to be launched to a neighboring star system sometime around 2060 –10/+20 years.

THE COMING SPACE AGE

The bubble chart below summarizes these forecasts more conveniently, allowing us to explore their significance. What is most striking is the prominent role of space tourism, standing out to the upper left of the chart. It is imminent, has high confidence, and is likely to transform space exploration.

A horde of robotics probes, planetary landings, and fly-bys will almost certainly continue to explore the solar system. One of Tech-

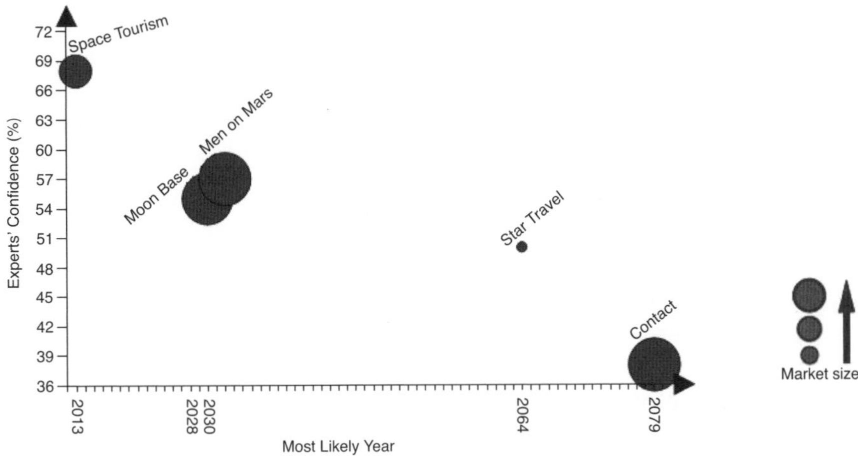

Figure 7.1 Bubble Chart of Space

Cast's experts, Peter King, thinks that the maturing of AI and robotics about 2020 should enable far more sophisticated unmanned missions. Building a Moon colony and landing astronauts on Mars also seem likely to occur in about 20 years, establishing beachheads beyond Earth.

The most seminal event could be the growth of a space tourism industry, offering vast opportunities for entrepreneurs to serve latent demands for space travel. Burt Rutan said: "If we reach our goal of flying 100,000 people in the first ten to twelve years of private space flight, you are going to see unusual creativity."[4] Supported by the genius of business people like Richard Branson of Virgin Galactic, space tourism is likely to signal a watershed event that shifts control of space to free enterprise.

When people realize they can travel into space at will, it is easy to imagine how attitudes could change dramatically. Space travel will no longer be a distant adventure reserved for astronauts, but a real part of ordinary life. Just as the Internet grew into the biggest technology system of the world in just a decade, space tourism could ignite the imagination of countless people with an adventuresome, entrepreneurial bent. Another TechCast expert, Jeff Krukin, Director of the Space Frontier Foundation, sees a new view emerging in which space activity becomes essential for creating a peaceful and prosperous world.[5]

Although NASA's central role may seem to contradict the trend toward private space ventures, it may be that the Moon and Mars will then become more easily available for private enterprise, much the way Army posts were established by the Federal Government to tame the west for a wave of pioneers. We could compare NASA's role to that of the Coast Guard in protecting vessels at sea, or the FAA in supervising air traffic. Other private projects, such as the space elevator, solar satellites, and the like could also encourage people to homestead on the Moon, Mars, and who knows what other bodies. And if communications were to be established with alien life sometime around this point, all bets are off.

The critical factor enabling a Space Age is likely to be the maturing of globalization into a fairly harmonious but diverse world governed as a coherent system, as we will see in the final chapter. Yes, I know this seems a fantasy, but all historic transformations seemed equally preposterous before they arrived. The fact is that globalization is one of the most powerful trends of our time, and the process has already caused the collapse of Communism, the rise of Asia, and a *de facto* global economy. Driven onward by the IT and knowledge revolutions, globalization seems likely to culminate into a global order of some type at about 2050, shortly after a Moon colony and Mars landing begin launching hordes of space pioneers into the solar system.

Making the next step to deep space travel will require fundamental breakthroughs in physics which we can not yet grasp. Travel to distant stars is so far beyond our present capabilities that it would be comparable to asking Columbus, Magellan, or the Vikings to travel to the Moon. They wouldn't have had an inkling – yet men landed on the Moon a few centuries later.

In a world of billions of educated people who have their needs fairly well met and are looking for further challenges, science will be in a position to take on deep space in earnest. Even now, basic concepts of physics are being overturned. Science is discovering how to control the speed of light, once thought by Einstein to be one of the fundamental constants of the universe. Under special conditions, light has been slowed, brought to a complete halt, and even reversed. Gravity – heretofore considered a mysterious force – is now thought to consist of waves and they have been measured. After all, something must be taking place between objects drawn together. How else would the Moon "know" the Earth is attracting it? The very nature of the universe is being explored, and the proven laws of quantum mechanics suggest an infinite number of parallel universes – a "multiverse." Atomic particles seem to flit in and out of existence, so they may possibly be moving between parallel dimensions of reality.

These are just tentative gropings, but as science advances with the collective knowledge of a unified world, they could form a different paradigm of physics that shatters what now seem impenetrable barriers of time and space. Astronaut Edgar Mitchell believes the Space Age requires a unified planet: "The exploration of deep space, of other planets and other solar systems, is such a tremendously expensive high-technology undertaking that a nation cannot do it alone... It' got to be a global effort."[6]

Using our analogy of the Earth as an organism, it could be said that this exodus will represent offspring of the planet sent out to perpetuate civilization throughout space, much as parents send their children into the world. And because space communities will possess the Earth's heritage of knowledge, just as a child inherits the legacy of its family, space exploration may initiate a new cycle of evolution at an even higher level. This could be the major step evolution has been building toward – launching humans throughout the inky depths of the universe.

Notes

1 Joel R. Primack and Nancy Ellen Abrams, *The View From the Center of the Universe* (NY: Riverhead/Penguin Books, 2006).
2 Steven Johnson, "The Long Zoom," *New York Times* (October 8, 2006).
3 Jack Hitt, "The Discover Interview with Burt Rutan," *Discover* (October 2007).
4 *Op. Cit.*
5 www.jeffkrukin.com
6 L. Fowler, "From Outer Space to Inner Odyssey," *New Age Journal* (1985).

Part II

Social Impacts of the Technology Revolution

CHAPTER 8

Shifting Structures of Society: Business, Government, and Other Institutions in a Knowledge Age

To appreciate the power of technology to transform society, recall the upheaval caused by the Industrial Revolution. Just two centuries ago, most nations were governed by the whim of kings, who were believed to derive their power from God. In the Western world, thought itself was carefully controlled by the Catholic Church. New technologies, economic markets, and more enlightened philosophies swept away this old feudal system, and replaced it with democratic government, corporations, labor unions, news media, global finance, and other social institutions we now take for granted. Today, something similar seems to be happening as the Technology Revolution transforms the structural foundations of society once again. A "creative destruction of social institutions" is underway.

This chapter begins the transition from Part I of this book, which provided forecasts of the Technology Revolution, to Part II, where we will focus on the social implications of this transition. The method in Chapters 8, 9, and 10 is quite different as we draw on social science rather than technology forecasting. Although the data may not be as rigorous, we can gain reasonably sound, intriguing insights into the crucial social issues of our time: the transformation of institutions, the emerging role of artificial intelligence (AI) versus humans, and scenarios sketching out the long-term trajectory of society.

THE CREATIVE DESTRUCTION OF INSTITUTIONS

This process of creative destruction can be seen in today's historic challenges to business, government, and other institutions. Corporations have taken the brunt of this upheaval in demands for social responsibility, a new breed of Gen X and Y employees who want to bring their dog to work, intense global competition, the energy and environment crises, and many other continual adjustments to the forces of change. The government bureaucracies invented decades ago have become anathema, although viable alternatives remain illusive. The American healthcare system is unsustainable, and public schools struggle with poor performance. Publishing, finance, and other information-intensive fields are being replaced by the Internet. A recent article in Fortune stated "Newspapers are dying."[1]

Beneath the surface of this confusion, however, the forces of technology grind on. Andy Grove, Chairman of Intel, put it best, "Technology always wins in the end." The collapse of the Communism, for instance, was not simply caused by agreements between Reagan and Gorbachev but by the raging forces of information technology (IT) that undermine authority everywhere. The USSR's planned economy could no longer cope with the complexity of the Information Age, and the seeping of modern media into Eastern Europe encouraged Poland, Hungary, and other satellites to revolt. Whether a planned economy, big government, or corporations, IT requires bottom-up systems like the Internet.

It was also technology that moved women out of the home to become professionals, executives, and politicians. Yes, the liberation of housewives was inspired by feminists, but it was only possible because contraceptives ("the pill") freed women from bearing an endless line of children, while automation shifted labor from factories to white collar work that women excel at. A few years ago it was inconceivable that women would become CEOs, much less president of the United States – which seemed entirely plausible as this book goes to press in late 2007.

There has been lots of minor organizational change over the past decades, but it has merely nibbled at the edges of a much larger problem. Talk to managers and employees about change today, and you are likely to be greeted with groans of frustration and rolling eyeballs. We have heard endless buzz over endless management fads yet seen so few serious results that organizational change is no longer taken seriously. A leading edge is always busy inventing new practices, but few managers have made a dent in the vast bureaucracies that continue to roll on. It's going to take something more fundamental to make working life productive and spiritually sane.

I have come to see the issue goes beyond ***organizational change*** altogether. The real problem is that entire classes of organizations (business, government, education, etc.) have become outmoded in this shifting environment, and so the big need is for ***institutional change.*** Unlike the relatively easy work of organizational change – process design, teamwork, leadership, etc. – institutional change involves redefining the underlying ***rules or values*** that govern these social structures.[2]

A prominent example is the deeply ingrained urge for hierarchical order. Most people are raised in hierarchical systems (family, school, church), so the idea is instinctively thought to be essential. We will soon see that bottom-up systems are emerging that are more responsive and creative. The traditional hierarchy is decentralizing into a self-organizing network of "**self-managed internal enterprises**" able to harness knowledge at the bottom of the organization to manage complexity.

The most striking example is the profit motive, which dominates corporations in much of the world. Profit served as the rightful business goal decades ago when our main task was to build an industrial economy. But this chapter will show that the old focus on capital is yielding to a new set of corporate values. We will see that there is an increasing emphasis on working relationships with employees, customers, and the public to form a broader form of governance that goes beyond the philanthropic idea of social responsibility. Collaborating with stakeholders can form a larger "**corporate community**" that improves the entire socio-economic system, including profit.[3]

Most of us take this vast infrastructure of society for granted because it is as ordinary and invisible as the air. Minor change takes place within these institutional boundaries, but challenging strongly-held institutional norms provokes confusion and resistance. Question the concept of hierarchy or the profit motive, for instance, and your most persuasive arguments will usually fall on deaf ears. These concepts are so deeply engrained in the cultural psyche that they seem inviolate, accepted as matters of faith, the only way the world is presumed to work.

Institutional change is a slow, agonizing process that requires yielding beliefs in revered but outmoded concepts. The following discussion will show that the Technology Revolution is driving a transformation of social structures, and the process may accelerate. If one grasps the newly emerging logic of knowledge, it is possible to understand the significance of what is happening and to anticipate how institutions will work in a knowledge-based world.

FROM HIERARCHY TO ENTERPRISE

The problem of hierarchy was nicely demonstrated in a corporation I worked with. The CEO was struggling with the typical bureaucracy that plagues most organizations. Her IT department exceeded its budget by leaps and bounds year after year, and the line units that used their services complained constantly about long delays, shoddy work, and unresponsive techies. She repeatedly attempted to control costs, ration usage, and replace managers, all to no avail. She was stumped for an explanation.

We introduced a different perspective of "internal market economics" that helped solve the problem, literally within days. Most managers have been raised on traditional hierarchical thought, so they don't realize that organizations are fundamentally economic systems. From a market economics view, the problem became quite clear.

On the demand side, IT costs were skyrocketing because the IT unit was a typical cost-center, subsidized by the CEO, so it was offering "*free goods.*" Line units could order IT services at no cost, so why not ask for more? On the supply-side, it also became clear that the IT unit was not performing well because it enjoyed a "*monopoly.*" Line units had no choice but to patronize this internal provider, and so the IT manager and his employees had little incentive to do a better job.

This analysis led to an incisive solution that was remarkable for its power and simplicity. *The CEO simply gave the IT budget to the line units.* Line units generated this money anyway, so the net effect was to return the "*taxes*" they paid in overhead to support this "*corporate government.*" The IT manager was invited to transform his unit into an "internal enterprise" – just like any small business. He was free to run it as he thought best and permitted to retain revenue exceeding costs. He liked the idea and found the challenge exciting. The line units were thrilled at having extra money they could use to either patronize the company's IT unit or to buy their services elsewhere.

With this shift in perspective, everything changed quickly. As some line units began patronizing outside providers, IT people began thinking seriously about how they could improve service to their new "*clients.*" And now that line managers had to pay with "*their own money,*" they began considering carefully whether they really needed all the extravagant stuff they once ordered. Service improved, costs declined, IT became a celebrated success, and the line units enjoyed their new found independence – all by *facing economic reality.*[4]

I think this illustrates the basic cause of today's bureaucratic ineffectiveness. Most managers are locked in a hierarchical view that ignores the economic realities driving their organizations. Studies show centralized management decreases the value of roughly half of all large companies. When the U.S. Federal Trade Commission was thinking of breaking up Microsoft, the value of its individual units was put at 30% above the stock price.[5]

In contrast, observe the robust growth of bottom-up systems as IT automates operations worldwide into distributed information networks operating in real-time. EBay is growing 60% per year because it formed an "internal market economy" that brings together sellers and buyers around the world, spawning an entire industry that didn't exit before. *MySpace* encourages its artist-members to sell their music and videos through the website. Cisco conducts 90% of all transactions with suppliers and clients using electronic systems. The speed of corporate decisions is so fast that "digital dashboards" provide real-time information to allow instantaneous organizational control, like a flight simulator. One CEO said it allows him to "feel the pulse of the business."

With organizations operating instantaneously *via* IT systems, it is no longer possible to supervise by standing over your employee's shoulder. In a wired world, leaders must concentrate on results, especially the overall performance of small units working anywhere in a far-flung knowledge economy. As much as 90% of the net worth of large corporations consists of intangible knowledge assets, such as skilled workers, partnerships, software, and the like. Capital remains important, of course, but strategic attention has moved to a laser-like focus on innovation, creativity, entrepreneurship, and other forms of knowledge work.

A good example is Napster, the music-swapping software system that gained 50 million members in weeks. The invention of Napster was so clever that it established a dramatically different peer-to-peer architecture. Yet the inventor was a high school student. BitTorrent was formed by another lone genius who designed a distributed system to circumvent the logjam of bits waiting to transmit movies on demand. The enormous success of Facebook was launched by two more students.

Similar examples abound in which people without advantage, resources, or status somehow gained the insight to produce creative innovations. Organizations are increasingly being driven from the bottom-up using principles of entrepreneurship to harness this talent. In a knowledge-centric world, we want to encourage ordinary people to introduce innovations and reward those who succeed. Yes, it will be messy, but innovation is a messy business. And with continued automation and the off-shoring of routine jobs unabated, the remaining tasks are entrepreneurial in nature anyway.

The growth of teams, performance pay, intrapreneurship, and other trends all point in the direction of ***self-managed internal enterprise***. A survey found two-thirds of professionals work in teams, and 90% think it is more productive and satisfying than the old system. Most corporations and many government agencies use incentives, bonuses, and stock plans to reward performance, often for entire teams. At IBM, 42% of employees do not have offices, and one-third of all AT&T managers are free to work where they choose. A manager at Best Buy said, "The focus is on whether we produce, not how, when, and where we do it." Allowing workers to become partners who take "ownership" of their units is the key to productivity and innovation. Some prominent exemplars are summarized below.[6]

Exemplars of Internal Enterprise

Best Buy saw 35% gains after moving to a "results only" system that allows employees to choose when, where, and how they work as long as they produce.

Amazon has 65,000 independent web entrepreneurs who sell its services on their own sites. These "mini-Amazons" create solutions while increasing company sales.

Johnson & Johnson has grown a robust 15% each year for 120 years by being decentralized into eighty small, self-managed businesses with an average of 350 employees. Each company has its own board.

Whole Foods uses small teams to manage each department in a store. Teams have authority to hire new members, control operations, and choose products, and the entire team is rewarded with bonuses based on performance.

Nucor is America's most successful steel firm because it ties pay to performance, with workers earning three times the industry average, about $100,000 per year. Here's how the CEO described absorbing failing steel mills into the company: "Once our culture is in place it far outperforms anything ... by 30%, 40%, 100%."

The City of Indianapolis was a government bureaucracy when Steve Goldsmith became mayor. After requiring units to compete with outside providers, costs and taxes fell, services improved, employee pay and job satisfaction rose, and an influx of people and business rejuvenated the City.

Semco thrives in its tough Brazilian economy because any employee can start a business. Workers set their own hours and choose their managers.

Google uses teams of three to ten people to manage each project. Teams operate like small internal ventures, making the company a sort of venture capital firm, placing bets on different internal projects.

Nokia uses autonomous business units to launch new phones each year. Even research operations are profit centers whose funding comes from line units, resolving the traditional conflict between R&D and Marketing.

Public schools are using vouchers, choice, and charter schools to offer alternatives to bureaucracy. One-quarter of all District of Columbia public schools are chartered schools, and half will be by 2014 if trends continue. Chicago's Mayor Daley said "We need competition in our schools; that's why our universities are the envy of the world."

FROM PROFIT TO COMMUNITY

I was privileged to witness a vivid demonstration of the power of collaboration when visiting a large company. Seated at a conference table were managers, labor leaders, suppliers, distributors, and even officials from the local government. What was most striking is that the president of the company did not seem an imposing person. He had no commanding presence, was clearly not a genius, and showed little charisma. How did he manage to pull this group of big egos together, I wondered?

As the meeting progressed, it became clear that this was a different type of leader. He was intent on encouraging the talents of the organization, and so he rarely spoke himself but was more intent on asking others for their views. What was most remarkable is that he really listened. Unlike most leaders, this humble man focused on understanding the reality of the situation. It was like a breath of fresh air! A leader who cares what people really think? Who wants to hear the messy truth? Who does not impose his solutions? Surely this was either a ruse or it didn't work, I thought.

But it did work. It energized the meeting. People brought out their problems, their ideas, their misunderstandings, and all the other hidden agendas we normally keep contained. The president simply asked an occasional question, made a few suggestions, and tried to clarify what was happening. Otherwise, the group controlled the meeting. Most importantly, the meeting affirmed that this was their organization; it was a corporate community of shared interests. They were responsible for its success or failure, so they did whatever was needed to make it work.

OK, this humble approach really works, but what about the leader, I worried? He was obviously not "in charge," and in fact he seemed a bit awkward and uncomfortable at times. Little wonder when people would complain about some aspect of the company or even criticize his behavior occasionally. How could he maintain his dignity, I wondered, much less the power needed to be effective?

Beneath this appearance of disregard was a deep sense of respect and affection. Not because he held the power of the president, but precisely for the opposite reason. He had voluntarily yielded his authority. Ironically, by giving up formal power, he was given far more real power. They would do things for this man no ordinary boss could even ask for.

This is but one example, yet I think it highlights a key principle of leadership today: ***in a world of escalating complexity and empowered people, leaders must cultivate the art of helping diverse groups collaborate.***

Corporations are the most powerful institutions in the world, yet they are also the most poorly understood. One of the most confusing features is the apparent conflict between profit versus the needs of employees, customers, the environment, the public, and other stakeholders. Business actually creates enormous social benefits, but the focus on money places executives in a self-serving posture opposed to democratic ideals of community and public welfare. American business, for instance, excels at creating wealth, but the social costs are huge. A focus on money is responsible for the largest gap between rich and poor since the Great Depression, a history of financial scandals like Enron, exorbitant CEO pay while thousands are laid off, the decline of employee benefits, and a backlash against globalization.[7]

The problem was manifest when GM and Ford plunged into crisis as rising gas prices halted the sale of trucks and SUVs in 2006. Ford lost $12.7 billion, the worst loss in its history, and GM lost $39 billion in one quarter – the third largest in the history of all business. By pushing profitable gas guzzlers and fighting with labor unions, the American auto industry has essentially been overtaken by Toyota's energy efficient cars and harmonious worker relations. Wal-Mart's reputation for poor employee benefits and cut-throat treatment of suppliers is thought to have reduced its stock value by $16 billion.[8] Airline employees agreed to accept large cuts in pay and benefits, but later found that the executives rewarded themselves with multi-million dollar pay packages; a mechanic at American Airlines called them "arrogant, greedy, and selfish," while a United Airlines flight attendant accused them of "squandering the company's goodwill."[9]

The concept of social responsibility has been cultivated to address this issue, and a few prominent firms have thrived under this philosophy. But the focus on *responsibility* has caused it to be seen as philanthropy, a luxury to be indulged in when affordable. The consensus of research studies shows only a very weak relationship between responsibility and profitability.[10]

Among business managers, the general attitude is that social responsibility is not to be taken very seriously. It is usually considered a sop to calm the public's dislike of big business, complied with when possible but easily abandoned in times of economic stress. Because the focus is on *responsibility* rather than *performance*, the concept is doomed to remain of marginal concern in a free market system. Models of social responsibility – the Body Shop and Ben & Jerry's – eventually floundered for various reasons.[11] Like it or not, business must compete to survive.

Figure 8.1 illustrates how the evolution of corporate governance has stalled in a conflict over the interests of capital versus those of society. The two contending models of business – the Profit-Centered Model and the Social Responsibility Model – are mirror opposites, pulling the enterprise in diametrically opposing directions.

I vividly recall the excitement when social responsibility was first introduced in the '60s. Not much has improved over these 40 years, reminiscent of other well-intended but futile struggles that were also conceptually misguided. The "war on drugs" comes to mind, as well as the "war on terror."[12]

A new model is emerging that reconciles these conflicts by viewing the enterprise as a ***corporate community***,[13] and others propose similar concepts for this new theory of the firm.[14] (Figure 8.1) It is now becoming clear that organizations are basically political in that investors, employees, clients, and other stakeholders are all power centers that are essential to success.

Managers today are dependent on educated employees to produce creative, high-quality work. Fierce competition has forced a constant

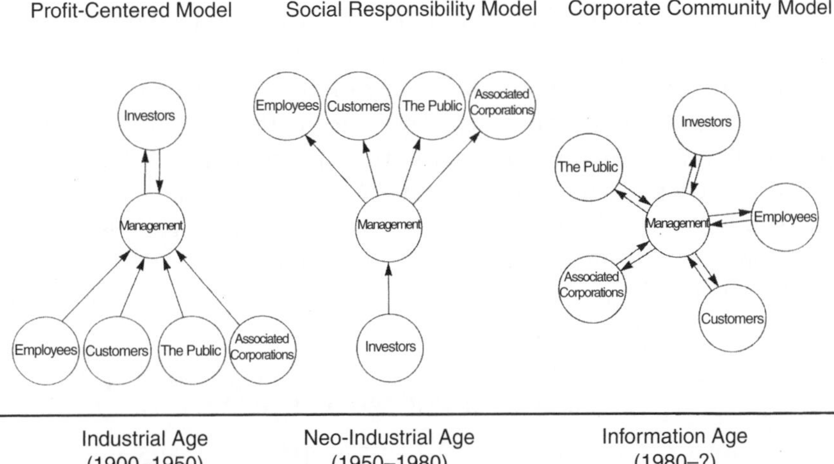

Figure 8.1 Evolution of Corporate Governance

drive to better understand customers' needs. Alliances with suppliers and business partners are now a competitive advantage. And the environment has become a strategic concern as developing nations increase the level of global pollution. Surveys we have conducted show 80% of managers know they must collaborate with their stakeholders.[15] But they are confused because this conflicts with the prevailing belief that the role of business is to make money. This conflict is solved by the logic of knowledge.

Unlike capital, knowledge can grow indefinitely. Capital consists of tangible assets (factories, land, money) that are limited, but knowledge is an intangible asset that *increases* when shared. Remember Ray Smith's "loaves and fishes" adage about knowledge from Chapter 3?[16] This means that collaboration is now economically productive because all parties can benefit through mutual sharing of knowledge. Michael Porter, the Harvard guru of strategy, notes "social responsibility has become an inescapable priority [and not] a zero-sum game. Both sides can benefit." A McKinsey study of 1,144 top global executives found 79% expect responsibility for social issues will fall on corporations, although only 3% said they do a good job presently.[17]

Figure 8.2 presents results of previous research modeling resource flows of the typical corporation. The data illustrate that social resources are several times as large as financial resources, confirming that social issues are actually more important. The data also show that all parties enjoy total net gains, highlighting the inherently productive role of business in creating social wealth as well as financial wealth. The main conclusion is that integrating these diverse resources into a more synergistic whole would create a far more productive and legitimate role for the large corporation.

The benefits of a collaborative approach are exemplified by the experience of Bonville Power, a utility company that struggled through the traditional adversarial relationship with labor unions, environmental groups, and local governments. They met with these constituencies, and after some heated exchanges, both sides found it possible to have a constructive dialogue. Here's what the CEO had to say: "Our adversaries helped us find creative solutions to intractable problems. Public involvement is a must today. Conflict is unavoidable; the only choice is whether to dodge it or harness it." Below are other examples of stakeholder collaboration:[18]

Exemplars of Corporate Community

The **Saturn** division of General Motors produces cars that rival Japanese competitors by engaging employees, unions, dealers, and local officials in managing the company. The loyalty of Saturn customers and employees is legendary.

IKEA has become a household name by providing low-priced, good quality furniture that clients can assemble themselves. Stores baby-sit kids, collaborate with suppliers, and boast egalitarian employee relations. The Company holds "Anti-Bureaucracy Weeks" where executives work as sales people and man registers.

The Container Store has expanded to 40 outlets and is growing 20% per year using a philosophy that places people before profits. Kip Tindell, the CEO, says "If you take care of employees, they'll take care of the customers – and that will take care of shareholders. To myopically focus on profits is wrong."

Whole Foods has multiplied its stock 25 times providing organic produce and a great place to work. Employees do high-fives in the aisles, and the CEO earns a paltry 14 times employee pay, compared with the typical multiple of 500. The CEO said "Profits are a by-product of treating people well, not the top priority."

Unilever has become a $52 billion global giant helping nations with poverty, clean water, food scarcity, and climate change. The CEO said: "You can't ignore your impact on the community. In the future, this will be the only way to do business."

The **Tata Group** in India is a tough global competitor with 32 businesses and collaborative employee relations. The company provides full healthcare, educational benefits, and its own schools and hospitals. CEO Ratan Tata says, unlike American managers, "Tata will not grow over dead bodies."

Nortel has strived to integrate profitability with responsibility. CEO David Ball reported studies showing that "stakeholder collaboration shows a direct correlation between financial results, customer satisfaction, and employee well-being."

Collaborative client relationships are productive A hospital taught its doctors to apologize when making mistakes, dramatically cutting malpractice suits. A car dealership allows mechanics to do whatever is needed to satisfy car owners; costs declined while business soared. A street vendor trusts his customers to make their own change, and finds that he gains their loyalty and can serve more people.

Novartis released its research data on drug development over the Internet, an unheard of idea in the high-stakes competition among Big Pharma. The company hopes to leverage the talents and insights of the global research community to speed up drug development.

The Model

Stakeholder	Resources (R)	Benefits (B)	Costs (C)	Return-On-Resources (B-C/R)
Investors	Equity/debt	Dividends/Interest Capital Gains	Capital Losses	Return-On-Investment
Employees	Education, Training, Knowledge	Wages & Benefits, Job Satisfaction	Disabilities, Meals & Travel, Job Dissatisfaction	Return-On-Human Resources
Customers	Search Costs, Purchase Price	Utility (Consumer Surplus)	Damages, Depreciation, Maintenance	Return-On-Purchase
Public	Public Assets	Taxes, Contributions	Public Services, Pollution, etc.	Return-On-Public Assets
Associated Firms	Assets of Firms	Sales of Firms	Expenses of Firms	Return-On-Associated Assets
Total Corporation	Total Resources	Total Benefits	Total Costs	Return-On-Resources

Typical Results ($ Thousands)

Stakeholder	Resources (R)	Benefits (B)	Costs (C)	Net Return (B-C)	Return-On-Resources (B-C/R)
Investors	$ 9,993	$583	$234	$349	3.5%
Employees	36,520	1,691	57	1,634	4.5
Customers	10,533	4,066	2,249	1,817	17.3
Public	2,536	338	375	–37	–1.5
Assoc'd. Firms	507	314	312	2	.4
Total Corp.	$60,089	$6,992	$3,227	$3,735	6.3%

Results from a computer simulation reported in Halal, "A Return-On-Resources Model of Corporate Performance," *California Management Review* (Summer 1977) Vol. XIX, No. 4.

Figure 8.2 The Socio-Economic Corporation

FROM CAPITALISM TO DEMOCRATIC ENTERPRISE

This has been a short analysis of a complex subject, so readers may want to explore the extensive literature on these difficult issues.[19] I think we have shown, however, how two overarching themes of self-management and corporate community are emerging, and that these concepts make sense economically. I know the attraction of hierarchy and money runs deep, it is very hard to change behavior, and most people do not understand these ideas. Little wonder it is such a struggle to alter institutions.

But change is being driven by the relentless pace of IT and the competitive race to use knowledge effectively. *BusinessWeek* noted that executive turnover increased by a factor of five during the past decade because "CEOs are barely in control of their companies' fate... It is increasingly difficult to digest the flood of information and make the right choices."[20] As the pace of organizational life approaches real-time, the present centralized form of corporate management may prove no match for the turmoil that is even now roiling modern economies.

The power of these two themes is that they represent an extension of our Western ideals into everyday life. Self-management brings the dynamic qualities of free enterprise inside organizations, and corporate community is the institutional equivalent of democracy. Both of these two principles emanate from the common need for collaboration, whether with employees, clients, or other stakeholders.

The mission of big business would evolve under this perspective into a "quasi-democratic institution" in which employees, customers, business partners, and the public join investors in advancing their collective success. But the purpose should ***not*** be an obligation to have "social interests represented," which reflects limits of the social responsibility view. The goal should be to ***enlist selected leaders of stakeholder groups as active partners in building a more successful business.*** CEOs are asked to accept responsibility for unifying the power of these diverse interests into a more productive whole. Surveys show that 70% of the public thinks corporations have excessive power and 95% think they should serve the interests of employees, their communities, and other groups in addition to money.[21]

I realize this is hard to do. Corporate management must invest time and goodwill to cultivate productive relationships with stakeholder groups, who may be difficult. Conversely, stakeholders have to provide active support rather than make demands. I hope this analysis has shown, however, that stakeholders are integral to any enterprise, only they know what they want, and they will assert their powers.

Collaborative problem-solving may be the most underutilized technique of our time.

This perspective could solve the nagging issue of political legitimacy that plagues big business, it could provide a more worthy corporate purpose, enhance economic performance, and serve as an exemplar for government, education, and other institutions to emulate. Economics has traditionally been called the "dismal science" because it presumed *limited* resources that *decrease* when shared to produce a world of *scarcity*. But this powerful idea introduces a world of *unlimited* resources that *increase* when shared to produce a world of *abundance*.

The biggest opportunity for change may be the political vacuum left by the second Bush Administration's policies. I mean no disrespect for the president, but the gap between rich and poor exceeds that of the Roaring Twenties, which ended in the Great Depression. The wealth of Bill Gates alone equals the total assets of 40% of all Americans, and the number of billionaires like Gates is skyrocketing. Meanwhile, middle-class incomes are flat, millions of jobs may be outsourced abroad, and social benefits are in decline. A survey of 178 nations found the mixed economies of Denmark, Finland, Norway, Sweden, and Switzerland had the happiest people in the world, and they were equally competitive with capitalist nations like the U.S. In fact Finland was rated the most competitive nation in the world.[22] Many think the 2006 national elections marked a popular uprising for these concerns of the middle class, and that we could see a revolt against the rich.[23]

What is most significant is the president's style of leadership, a topic that I have studied extensively. In an age when it is common to see leaders trying to woo their constituencies, George W. Bush stood out as an unapologetic autocrat. In his own terms, he was the "decider." This is a courageous stand, but it fails to meet the needs of a more complex world with discriminating people. The conflict in Iraq, for instance, demonstrates how sheer military might and lots of money have very limited impact in today's smaller world. Conversely, Iraq also suggests the unavoidable need for collaborative problem-solving, community-building, and simply understanding other cultures. These qualities increasingly form the heart of modern democracy.

American politics moves in roughly a 20-year cycle, and severe disappointment in the second Bush Administration seems to have ended the conservative wave that began with Ronald Reagan's election. Progressive leaders could find rare opportunities to fill this vacuum, but there will be little tolerance for the big government of the past, which has also been discredited. This failure of the Left and Right opens up vast possibilities for what Mark Satin called the "radical middle."[24] That is precisely the potential of the present trends we have described that are bringing free enterprise and democracy inside institutions. Collaborative working relations and community serve the public welfare espoused by liberals while internal enterprise enhances the self-reliance and market forces espoused by conservatives.

For instance, this combination of left- and right-wing policies could resolve the dilemma facing American healthcare as costs approach 20% of GDP. U.S. healthcare costs almost twice as much as other industrialized nations yet provides mediocre results by almost any measure, and the U.S. is alone in the modern world by having 46 million uninsured people. This situation is unworthy of a great nation.[25] Yet conservatives want to continue letting the free market solve the problem, while liberals want a government-paid system. To believe more market freedom will solve this complex dilemma is magical thinking, and a government funded system smacks of socialism to most Americans.

The outlines of a likely solution are emerging now, and prospects look promising because the emerging model is a synthesis of government support and market forces. Various parties seem to be moving toward a consensus on the following features of a new American healthcare system:[26]

The Emerging Consensus on American HealthCare

Universal Insurance Coverage The Federal Government would require all to have basic healthcare insurance, and it might organize "exchanges" through which people can select among competing plans. The poor would be offered free vouchers good for basic health coverage, while rich may be able to opt out by being self-insured.

Employers Relieved of Responsibility Corporations and other employers would be freed of the responsibility for healthcare. Business could then become more competitive by avoiding the $500 billion they now spent annually on health insurance.

Providers Evaluated on Results One of the great flaws in the present system is that there is little or no information to help make sound decisions. But plans are underway to require hospitals and physicians to be evaluated for providing results. Patients could then make wiser choices and thereby allow market forces to improve the system.

Minimal Added Cost or Bureaucracy This solution would simply shift costs from employers to individuals, resulting in little added cost or federal programs. The costs of vouchers for the poor could be offset by higher tax revenue as corporations are better able to drive robust growth and market forces improve efficiency of the system.

With similar imagination, political skill, and this same blend of left-wing social action and right-wing self-reliance, we could conceivably put to rest the nagging arguments over guns, abortion, terrorism, and other social dilemmas. Why can't Americans accept the freedom to own guns but also require registration, training, and safeguards? Maintain the right to abortion but also encourage alternatives? Tighten worldwide efforts to arrest terrorists while also supporting the economic development of Muslim nations. Yes, these are political minefields, but where is it written that the U.S. can not match Europeans in social progress?

Enlightened business executives should take the lead by redefining the corporation as a community that provides economic gains while also serving society. A bold move like this would allow business leaders to assume a more honorable role as stewards of today's socio-economic corporations. And because institutions are interconnected, there is an intriguing possibility that government could relinquish much of its role overseeing business because firms would become self-regulated. The economic system that results would no longer be "capitalism" but a more vibrant, harmonious market economy I call "democratic enterprise."[27] Redefining institutions – especially the corporation – could be the single most powerful move to improve life in the difficult years ahead.

This may seem idealistic, but these principles have withstood the test of history precisely because they use knowledge effectively. The genius of free enterprise is that markets draw on the creative knowledge of entrepreneurs and the discriminating knowledge of customers to improve the economic system. And democracy relies on education, debate, and knowledge to reach balanced choices that are legitimate. It is no coincidence that great corporations are guided by roughly this same philosophy.[28] Without much planning or forethought, these two pillars of enterprise and democracy are being extended inside the interlocking maze of institutions that form the structure of society itself. The examples above represent an *avant garde* that has experimented with such innovations for decades. Now we simply have to bring it into the mainstream.

It has always struck me as odd that Americans are so proud of their heritage of markets and democracy – yet they accept the fact that daily life is governed by the planned economies of large institutions and the benign autocrats who rule them. It may be that American culture is too strongly wedded to a pure form of capitalism to make these difficult changes. Perhaps the leadership will come from Europe or Asia. I think we will learn the outcome in the next decade or so as nations strive to fill the current vacuum in political leadership. Imagine what could be achieved if we embraced these ideals throughout society.

Notes

1 Marc Gunther, "Hard News," *Fortune* (August 6, 2007).
2 Douglas North, *Institutions, Institutional Change, and Economic Performance* (Cambridge Press, 1990).
3 Halal, "The Collaborative Enterprise: A Stakeholder Model Uniting Profitability and Responsibility," *Journal of Corporate Citizenship* (2001) Issue 2.
4 Reprinted from my article in Bill Treasurer (ed.) *Positively M.A.D.* (SF: Berrett-Koehler, 2004).
5 "Too Much Corporate Power?," *BusinessWeek* (September 11, 2000).
6 Examples are from "The Net's Next Phase," *Fortune* (November 13, 2007); "Break Free," *Fortune* (October 1, 2007); "A Boom for DC Charter Schools," *Washington Post* (February 25, 2007); "The Future of the Workplace," *ABC News* (September 4, 2007); and my previous book, *Internal Markets* (Wiley, 1983).
7 Robert Samuelson, "The Rich and the Rest of Us," *Washington Post* (April 18, 2007).
8 "Beyond the Green Corporation," *BusinessWeek Online* (January 29, 2007). "Why Wal-Mart Can't Find Happiness in Japan," *Fortune* (August 6, 2007).
9 Del Quentin Wilber, "Turbulence Over Executive Pay," *Washington Post* (May 22, 2007).
10 Harrison and Freeman, "Stakeholders, Social Responsibility, and Performance," *Academy of Management Review* (1999) 42:5.
11 David Vogel, "Is There a Market for Virtue?," *CalBusiness* (Winter 2006).
12 Zbigniew Brzezinski, "Terrorized by the 'War on Terror'," *Washington Post* (March 25, 2007).
13 See Halal, "The Collaborative Enterprise," *Op. Cit.*
14 Porter calls this "corporate social integration" and Laura Tyson thinks of it as "social entrepreneurship." *BusinessWeek* (May 3, 2004).
15 "The Collaborative Enterprise," *Op. Cit.*
16 Halal, *The Infinite Resource: Creating the Knowledge Enterprise* (SF: Jossey-Bass, 1998).
17 Michael Porter and Mark Kramer, "Strategy & Society," *Harvard Business Review* (December 2006).
18 These examples are from Thomas Friedman, "The Whole World is Watching," *New York Times* (June 27, 2007); Don Tapscott and Anthony Williams, "The New Science of Sharing," *Businessweek.com* (March 2, 2007); Simon Zadek, *The Civil Corporation* (London: Earthscan, 2007); and my book *The New Management:* (SF: Berrett-Koehler, 1998).
19 In addition to my own works cited here, see Gary Hamel, *Leading the Revolution* (Boston: Harvard Business School, 2002).
20 "The CEO Trap," *BusinessWeek* (December 11, 2000).
21 "Too Much Corporate Power?," *BusinessWeek* (September 11, 2000).
22 "Shiny, Happy people," *Discover* (December 2006); Jeffrey Sachs, "Welfare States, Beyond Ideology," *Scientific American* (November 2006).
23 "David Broder, Independence Days," *Washington Post* (September 21, 2006).
24 See Robert Olson, "The Rise of Radical Middle Politics," *The Futurist* (February 2005).
25 For a complete analysis of American healthcare, see "World's Best Medical Care?," *The New York Times* (August 12, 2007).

26 See Michael Porter and Elizabeth Olmsted Teisberg, *Redefining Health Care* (Boston: Harvard Business School Press, 2006), David Broder, "A Market Makeover for Health Insurance," *Washington Post* (October 14, 2007).
27 Halal and Taylor, *21st Century Economics* (NY: St. Martin's Press, 1998).
28 Jerry Porras and Jim Collins, *Built to Last: Successful Habits of Visionary Companies* (NY: HarperCollins, 2002).

CHAPTER 9

An Age of Consciousness[1]: The Next Place in Technology's Promise?

The enormity of the question of "consciousness" was unexpectedly dropped into my lap when flying home to Washington, DC, from a trip to Korea. My row companion was a Korean man who left a life of financial success to become a Baptist pastor in the DC Area. I told him who I was, and almost immediately he looked directly into my eyes and asked "Are you saved?" "Have you accepted Jesus Christ as your Savior?" He persisted in devoutly explaining "the path to salvation," but later we were able to exchange views a bit, and I think he came to agree there are *many* paths to salvation.

This ordinary exchange nicely captures the wondrous diversity of countless other ***islands of consciousness*** that meet in work, families, politics, and all social functions. Life is lived moment to moment as our individual points of view intersect, sometimes productively and other times destructively. Consciousness pervades life – yet we have only a vague inkling of what it really is.

This chapter argues that consciousness is the next great frontier in civilization's progress because it exists beyond information, knowledge, and other forms of "rational" logic. The world of spirituality, ideology, culture, values, goals, purpose, choice, and many other aspects of subjective experience may soon take center stage for the same reason the Information Age emerged and the Industrial Age before that – because a shift in consciousness is needed to solve the challenges ahead. Previous chapters showed that technology's promise requires containing a five-fold increase in industrialization, a commensurate leap in environmental stress, conflict among nations armed with weapons of mass destruction (WMD), the revolutionary powers of bio/nano/AI technologies, and other challenges that can not be resolved within our existing state of mind.

Driving us in this direction is the relentless pace of information technology (IT), which is likely to pose an existential dilemma. Intelligent systems are encroaching on us with androids replacing humans, intelligent agents acting as our assistants, smart computers that talk, discoveries in neuroscience able to model the brain, and endless other forms of artificial intelligence (AI). This "automation of mental work" poses one of the most intriguing scientific issues of our time – ***Is there a fundamental difference between machine intelligence and human intelligence?*** Despite the fact that most people are convinced that human thought surpasses sheer information, could we all be wrong? After all, everybody accepted the Flat Earth view for millennia.

The next section outlines why consciousness poses such a perplexing scientific challenge. Then I summarize evidence arguing that it is a physical property of the brain and opposing evidence suggesting that it arises out of some higher form of "human spirit" that is not well-understood. We show that a synthesis of these views presents a far richer and more powerful perspective, and that "technologies of consciousness" are likely to enter mainstream use about 2020 +/– 5 years – out of sheer necessity.

WHAT IS CONSCIOUSNESS?

Consciousness is such a common part of life – like breathing air – that it is taken for granted. But with the advance of IT, this field of study has exploded into a central scientific debate. To avoid getting lost in abstractions, let's start by looking at examples of consciousness in action. Here are some personal accounts of people experiencing various shifts in mental states,[2] often dramatically:

Vignettes of Consciousness

A young man meditating "Slowly, the mind quiets. I float. All my energy seems to draw into a ball moving a foot or so above the floor. Then the ball bursts, turns liquid, and rises upward, becomes pure fluid energy. Up and up until I erupt into an open space filled with music and joy and love. The sweetness of it is exquisite, almost unbearable."

A woman crying over loss in her family "I suddenly felt like a girl again after my parents' divorce. All at once, I was crying. It was as if I had an infected area where I had been stuffing my loss. I grieved openly and unashamedly for my mother and father, and for myself as a wife and a mother. This went on for 45 minutes. When it was over, I felt complete, cleansed, healed."

A young girl dancing "I didn't know him, and we started out tentatively. But soon we were making great loops around the room. It seemed that we weren't dancing at all; we were flying. We had risen a few inches off the floor, and we existed somewhere outside the law of gravity. The music, the other couples, and all else faded from thought as I danced in perfect joy and grace."

A soldier's experience of war "It is only now that I realize what a frightening experience I had been through. I was so completely wound up and braced for war that everything was taken in stride. Fear and disaster can be faced in the heat of battle if one subdues the emotions. But when it is over, then it is possible to emerge with full consciousness of the real world."

A woman's discovery of faith "I had the sense that something spiritual was coming into my life. There was no doubt, everything had shifted. I attended a Quaker meeting, and when I walked into that peaceful room, the silence spoke to me. Quakers believe that God is Inner Light that we seek together. The silence is so profound that sometimes we all become one with God."

These are vastly different experiences but they all reflect heightened states of consciousness. The high of meditation, the cleansing of grief, the flight of dance, the terror of war, and the discovery of faith go beyond rational logic – they are "higher-order" states of consciousness. Later we will see that this constitutes what I think of as "human spirit." There may be controversy over this phenomenon, but consciousness is very real and the central issue in life. The big questions, however, are what causes it and how does it work?

I am not a neuroscientist, psychologist, or otherwise equipped for original research in this complex field, and so I will let these authorities provide precise definitions of the mind, awareness, emotion, and other aspects of this large and varied phenomenon. I am an experienced scholar, however, and I think it's possible to pull together the theories and evidence into a coherent whole, focusing especially on the question posed at the start of this chapter: Is there a fundamental difference between higher-order human thought and sophisticated AI? In other words, can the essence of human consciousness be reduced to machines, or is something else involved?

Let's begin by summarizing how the mind works in broad terms. Neuroscientists have been probing the brains of various creatures with electrodes, and they find that the brain focuses on targets of interest. A hostile animal, a bit of food, or anything else can cause a greater number of brain cells to fire as attention is directed where it is needed. Out of the maze of sounds, sights, and smells that bombard the senses, animals ignore most signals and try to better understand particular targets. They then draw on memory, instinct, or learned rules to interpret what is being observed and take action. Simple actions may be automatic, but the most crucial function of the mind is essentially decision-making.[3]

This much is fairly clear, but the trouble occurs when trying to explain how a decision is actually accomplished. How do creatures direct attention? Make choices? Select goals? In other words, who is doing the deciding? To highlight the two major theories on this point, I describe below the dominant scientific view that consciousness arises from the brain – what I will call the "computer model." I then examine the view that higher consciousness involves some form of autonomous human spirit – the "spiritual model." We'll see that both perspectives have support.

The Computer Model

Scientists are usually "materialists" because of the nature of empirical research, so most attribute consciousness to the information systems of the brain – the computer model. From this view, the brain processes sensory information, draws on memory and other knowledge, and applies rules to make decisions, much as a computer would. The more rarified aspects above – choice, free will, goals, values, etc. – are simply higher-order functions that *emerge* out of the more basic information processes, so consciousness is determined by physical causes. Thought may be exceedingly complex from this view, but everything we experience can in principle be explained as a direct outcome of our "wet computer." This logic is appealing, but it reduces humans to computers, raising the prospect that machines will one day eclipse flesh and blood.

A prominent perspective in this vein draws on probability, statistics, economics, game theory, and the Nash Equilibrium that won John Nash a Nobel Prize. This rational view claims any organism can be modeled as a decision-maker driven by Darwinian logic to find optimal solutions to complex problems involving uncertainty.

A wonderful example is offered by studies of how ducks respond to bread balls thrown into a lake. One experimenter throws bread balls weighing 2 grams every 5 seconds while the other experimenter 20 feet apart throws the same bread balls every 10 seconds. Using the Nash Equilibrium, the optimal location is calculated to define how ducks can maximize their catch. It turns out that this occurs when two-thirds of the ducks place themselves near the researcher throwing bread balls every 5 seconds and one-third near the one throwing every 10 seconds.

Results showed an uncanny approximation to this optimal solution. Within 60 seconds, two-thirds of the ducks moved to the spot where food was dropped every 5 seconds and one-third went to where they were fed every 10 seconds. Furthermore, when the relative size of the bread balls and their frequency was changed, the ducks immediately repositioned themselves in accordance with the theory. Experiments with monkeys, birds, and mice bear out the same conclusions. In the words of neuroscientist Paul Glincher: "Animals come remarkably close to achieving optimal solutions. Mind simply does not figure in the equation."[4]

Not surprisingly, scientists are largely united in agreeing such logic can explain away the "ghost in the machine." Daniel Wegner, a psychologist at Harvard, claims the concept of self, will, or soul normally associated with personal identity is an illusion. Consider the work of a few well-known authorities:[5]

Support for the Computer Model

The Genetics Roots of Behavior Pulitzer Prize-Winning Scientist Edward Wilson founded the science of sociobiology, which explains behavior as inherited genetic traits. Wilson said: "Scientific knowledge holds that religious experience is entirely neurobiological."

Religiosity in the Brain Studies demonstrate that spiritual experiences are traceable by PET scans to particular regions of the brain. Michael Persinger, a neuroscientist, concludes "Religion is a property of the brain and has little to do with what's out there."

Decisions are an Illusion Benjamin Libet, a University of California physiologist, finds a one-third second delay between the time an act is initiated in the brain and the time the subject reports a decision to act, suggesting that conscious decisions are a mere afterthought.

The Power of Chemistry Rick Strassman, a psychiatrist at the University of New Mexico, finds that a compound called DMT is produced by the brain to create mystical visions and psychotic hallucinations.

Intuition is Ordinary David Myers, psychologist at Hope College, thinks intuition is not a mystical sense but merely a subtle form of learning that is often wrong. Basketball players who think they are on a "hot streak," for instance, score no higher than they would ordinarily.

Emotion a Form of Thought Antonio Damasio, a psychiatrist, finds that emotion is a form of reasoning performed by the body rather than some form of ethereal experience. "The division between reason and passion, or cognition and emotion, is from a neurological view, a fallacy."

Out-of-Body Phenomena Explained Olaf Blake, a neurologist, reports that mild electric currents to the brain cause out-of-body experiences. One subject said when the current flowed "I am at the ceiling," and when it stopped "I'm back at the table now. What happened?"

This is compelling evidence that behavior is governed by rational logic, genetic inheritance, brain chemistry, emotions, and a host of other physical factors. Nobel Laureate Francis Crick, Co-discoverer of DNA, carries this logic to its conclusion: "Your joys, your sorrows, your sense of personal identity and free will, are no more than a vast assembly of nerve cells and their associated molecules."[6]

Some scientists find this unconvincing. Antonio Damasio, the psychiatrist noted above who contests the ethereal nature of emotion, acknowledged, "Every part of the organism, every cell, is not just animated but living. Life is a strange state and no machine is equivalent to the living cell."[7] David Gelernter, professor of computer science at Yale, sees a similar flaw that violates Emmanuel Kant's distinction between subject and object. "All we can discuss scientifically is objective knowledge," says Gelernter. "Consciousness is subjective ... [a machine] has no inner mental life, no 'I,' no 'sense of self.'"[8]

That is the mystery. Where does subjectivity and that vibrant sense of self come from? There must be a decision-maker selecting things to focus on, choosing goals, and exercising free will to make deliberate choices. In surveys with audiences, I find that about 90% agree "There is a fundamental difference between humans and computers." They think there is something about the human mind that will always make us superior to the most intelligent artificial system. People may not be able to put their finger on it and they could be proven wrong. But the computer model does not seem to coincide with everyday experience.

The Spiritual Model

These difficulties lead to the spiritual model of consciousness. This does not imply religions and supernatural beings but simply recognizes that there exists a higher-order state of mind or "human spirit" in the sense defined by Webster's dictionary: "Will, consciousness, frame of mind, disposition, mood, as 'In high spirits.'" Think of the vignettes of meditation, grief, war, dance, prayer we saw earlier that portray the fluid, ethereal shifting of human spirit.

Or consider the ebb and flow of your mind during a typical day. You wake and leave the dreamy haze of sleep, then take a shower to refresh the senses, use coffee to stimulate attention for work, spring to battle over a challenge, enjoy the relaxation of a drink before dinner, and finally are carried off into sleep again. In short, life's daily cycle is a sequence of small but important changes in consciousness.

This prosaic human spirit is affected by a host of forces that shape our moods: social life, major news events, intellectual ideas, the high of alcohol and drugs, meditation and prayer, psychotherapy, social rituals and ceremonies, art and music, dreams, the influence of weather and seasons, physical activity like jogging, bodily rhythms, and almost anything that can affect the mind. Later, we'll present trends showing increasing use of these "technologies of consciousness."

The Dalai Lama has made a point of integrating his long tradition of studies in consciousness with modern neuroscience and argues that this "luminous nature of awareness" transcends the brain.[9] Daniel Batson, a University of Kansas psychologist, said, "The brain is the hardware through which [consciousness] is experienced. To say the brain produces consciousness is like saying a piano produces music."[10] There is a long and growing tradition of explaining the human spirit as a distinctive phenomenon acting under our control with profound consequences.[11]

Most adults are preoccupied with such subjective matters that defy logic because they transcend formal knowledge. The crucial actions of life – succeeding in a career, keeping a marriage alive, raising children, etc. – require drawing on mental strength, wisdom, and willpower. Tell executives, parents, the devout, and others that their actions are an automatic response to environmental stimuli and you are likely to get a blank stare. To most people there is no need for proof – free will and human spirit are self-evident. Consider the evidence on the spiritual side of this cosmic debate:[12]

Support for the Spiritual Model

Health Affected by Mind Hundreds of medical studies show people who feel in control of life, connected to others, and optimistic tend to be healthier and live longer. The well-known placebo effect is so strong that medical trials control for it, and the mere belief in a sugar pill can be more effective than the best medications.

The Power of Religion Those practicing a religion live longer, have less depression, and are generally healthier, even after controlling for smoking, drinking, and exercise. Dr. Jeff Levin at the National Institutes of Health reviewed the studies and found "The weight of evidence overwhelmingly confirms that spiritual life influences health."

Spirituality and Community Studies in scientific journals demonstrate a link between meditation and social order. The ratio of people meditating seems to be inversely correlated with crime and violence and directly related to positive behavior. In Christianity, the role of cloistered nuns and priests is to save the world through prayer.

Parapsychology Shows Significant Results A recent survey of extrasensory perception concluded the evidence is as strong as medical studies but scientific bias prevents acceptance. Meta-analyses of the literature show significant results and concluded that the beliefs of researchers are critical to results, the "experimenter effect."

The Source of Creativity Creative people are often mystified by the source of their inspiration. J.K. Rowling, author of the Harry Potter books, said, "It feels as if someone zapped the ideas into my head." Playwright Michael Frayn described it this way: "It came into my head instantaneously. I remember the moment very clearly," and novelist Saul Bellow said: "The book just came to me. All I had to do was catch it in buckets."

The Nobel Effect Studies show Nobel Laureates live on average two years longer than their peers who do not get the award, and similar studies show that Oscar winners also live longer. This effect is attributed by the researchers not to money or other physical factors, but to the supportive influence of being recognized as outstanding by society.

Does Universal Energy Shape Life? Rupert Sheldrake, a biologist at Cambridge University, has conducted experiments suggesting that invisible energy fields organize life. While this is controversial, studies show that learning occurs more easily after others have mastered a task, that people can sense being stared at, and dogs know when their masters are unexpectedly coming home.

Near-Death Experiences Untold numbers of people who were thought to be dead and returned to life relate the same experience of being greeted by loved ones and religious figures. One man told of seeing intense light like "A mother's love only a million times stronger." Afterward, they have no fear of death. Bruce Greyson, a psychiatrist, said, "Brain chemistry does not explain this."

The data above does not prove much, but they suggest that human spirit exerts subtle but powerful influences. Health is affected by beliefs, friendships, and optimism. Prayer, meditation, and other spiritual practices seem to improve personal well-being and communities. There are even some indications of telepathic communication, universal life forces, the special nature of creativity, and life after death experiences, although many doubt the studies are valid.

The human spirit is somewhat arbitrary and can change dramatically when people go through various transformative processes – scientific revolutions, nervous breakdowns, falling in love, religious conversions, and other shifts in highly abstract beliefs. The President of the United States himself, George W. Bush, is the product of a spiritual transformation, and Russia's collective soul went through a transformation when Marxist beliefs were overthrown for free markets and democracy. These basic changes in ideological frameworks, values, and beliefs are especially powerful because this higher level of thought controls lower-order functions of the mind.

This higher-order consciousness can be thought of as spiritual energy, constantly adapting in the struggle to find a trustworthy island of consciousness able to see us through the sea of life. Each organism is imbued somehow with this energy, and these islands of consciousness aggregate into larger islands as groups, organizations, and entire national cultures develop a spirit all their own. David Chalmers at the Australian National University said, "We are likely to discover that consciousness is a fundamental property of the universe, like space, time, and gravity."[13]

Although this is an appealing idea, it also raises the unavoidable question, "If there exists a domain of human spirit beyond information and knowledge, where does *that* come from?" There is no satisfying answer to this basic mystery of consciousness. The "strong" version of the spiritual model holds that all individuals are connected through some form of spiritual energy that unites the universe, which is the fundamental belief system underlying all religions.[14] Asian philosophers call it "chi," Western philosophers thought of it as "the life force," and for the Catholic Church it is "grace." As the evidence above suggests, this may be an entirely logical cosmology. This "universal" spiritual energy may have its own laws and logic and we simply don't understand them.[15]

It's tempting to believe in vast powers of the mind, but the issue is far from settled as many of the above studies produce conflicting results. True believers were humbled years ago when massive psychic energy failed to save the Skylab satellite from plunging to Earth by altering its orbit. Despite millions of well-intended souls focusing their thoughts skyward from the U.S., Canada, Great Britain, and Australia – the satellite didn't budge.

A Synthesis of Science and Spirit

This reminds us that nothing beats reality. The universe is a very complicated place, and these opposing views are engaged in a great scientific experiment now underway to establish the nature of consciousness.

Science is making enormous contributions by demonstrating that important aspects of behavior can be modeled as physical, deterministic systems, possibly leading to another historic revolution. Just as Galileo shattered conviction in an Earth-centered universe and Darwin dispelled the distinction between humans and animals, neuroscience is challenging deeply held beliefs about the mind. But without free will, how could we punish crime? Reward the successful? Aspire to betterment?

It is equally clear that *all* behavior may not be attributable to brain functions. The evidence also suggests that beliefs, spiritual practices, social relationships, and a host of other forms of human spirit also play a crucial role. The spiritual model could gain ascendancy as it becomes increasingly clear that good health, strong communities, global order, and other aspects of life can be explained and cultivated as higher-order mental functions. Widely recognizing these powers of human spirit could prove as revolutionary as the scientific revolution posed above.

I suggest these two theories can be integrated to support one another quite nicely: ***accept the considerable influence of the brain on consciousness, but also accept the considerable role of a higher-order human spirit.*** Freeman Dyson, a famous scientist, said: "Speaking as a physicist, scientific materialism and religious transcendentalism are not incompatible. We have learned that matter is weird. It does not limit God's freedom."

This synthesis of science and spirit could also resolve the endless haggling over evolution. If scientists and creationists would stop shouting at each other, we might see that a richer view of life is possible that reconciles both views. The evidence supporting evolution is beyond dispute, but it merely explains the physical mechanisms involved. A higher form of intelligence may guide evolution as many claim, but it could be the intelligence arising out of human sprit (and possibly "animal spirit") rather than a biblical act of creation. We may not understand the source of consciousness, yet the transcendent nature of spirit may explain why the world evolves at all. Life seems to be energized and guided by all those creative ideas we pluck out of thin air, values and beliefs, and the tough choices that shape conscious acts. Endless little sparks of spirit may guide life to myriad small steps that collectively determine the course of evolution.

I understand some find this troubling, but this tentative conclusion is based on studies of observable facts. Yes, consciousness depends on our marvelous biological machinery, but that does not mean it is reducible to the machinery. The challenge now is to reconcile these two models by exploring the limits of each and how they interact.

We know so little about spirituality that it is often dismissed as ignorance or fantasy, but that's because its very nature transcends rational logic. Human spirit seems to involve "belief systems" that organizes information into a coherent whole, much like the operating system of a computer. Belief systems act as an interface between the subjectivity of the user and objectivity of the machine. They are essential to make sense of a complex environment. Without beliefs, there is no way to organize the zillions of bits of information we rely on. Conversely, changing beliefs reorganizes this information into more useful forms. The philosopher Emmanuel Kant thought "will" and "idea" form the basis of reality, while the Buddha said, "With our thoughts, we make the world." A meeting of scientists puzzling over the erratic results of parapsychology studies concluded, "Some phenomena depend on beliefs."[16]

The pervasive nature of religion is especially striking. All societies throughout history have cultivated religions and ideologies because some type of belief system – even belief in atheism – is essential to make sense out of the maze of information flooding the brain, giving life purpose, guidance, and meaning. Even now in high-tech America, surveys consistently report that roughly 90% of Americans believe in a supreme being with whom they pray or talk.[17] One may not like the "religious right" in America, but it clearly plays a vital role in shaping a coherent inner world of family and community. Europeans seem to have become non-religious, but hundreds of new religions are forming in formerly Marxist Russia and China to satisfy the universal need for spiritual union.[18] Spiritual behavior is the most prominent and universal feature of life because it addresses the subjective realities flesh is heir to.[19]

The recent spate of books condemning religions rightly point out the illogical, destructive nature of religious dogma, especially when it inspires hatred, conflict, and illogical beliefs.[20] This questioning of outmoded religious ideologies is refreshing and spreading widely. But I hope our analysis illustrates that some type of belief system is unavoidable. Our goal should be to create more sophisticated belief systems that accord with the complexity of modern life.

Although the spiritual dimension is self-evident to most, it eludes a large number of people, especially scientists fixated on a material view. Richard Dawkins, the biologist who achieved fame denying religiosity, participated in an experiment known to produce profound spiritual experiences. Dawkins reports he "didn't feel a thing." Contrast this with the experience of another scientist, Paul Ekman, upon meeting the Dalai Lama. "I was inexplicably suffused with a wonderful warmth throughout my body. It was palpable. I felt a goodness I'd never felt before in my life."[21]

I think this may explain why the scientific and spiritual views are held in such strong opposition to one another. Because spirit is by definition a higher level of consciousness, it can comprehend lower level phenomena but the reverse is not true. For instance, we have the power to understand the intricate workings of that wondrous colony of cells we call our bodies – but they have no conception of the larger system they comprise. Life seems to be stratified into a hierarchy ascending from physical matter, to social behavior, to intelligence, and finally spirit, which controls the entire hierarchy from the top-down.[22]

The self-control vested at higher levels in this hierarchy accounts for the freedom humans enjoy. Even after recognizing the coming ability of sophisticated AI to learn, adapt to change, simulate emotions, and the like, somebody has to provide the free will to act. Nobel Laureate Roger Sperry thinks the mind is an independent force[23] and other neuroscientists find that the enormous plasticity of the brain allows us to gain control over life.[24] Ilchi Lee, the founder of a Korean Yoga movement spreading around the world, put it this way: "The truly important thing is not knowing but doing. We are what we do. Through the process of choosing, acting, and reflecting, our thoughts, actions, and habits evolve."[25]

The life of John Nash himself attests to the central role of free will. As his story is told in the movie "A Beautiful Mind," Nash was incapacitated for years by paranoid delusions. Through sheer will he finally mustered up the conviction to learn that he could ignore these delusional impulses, and thereby regained control of his behavior to receive the Nobel Prize for his path-breaking work.

With such evidence, we should at least recognize the mere possibility that a higher domain of consciousness could transcend the physical world. For instance, discoveries about the uncertainty of matter at the quantum level, entanglement of atoms, the Big Bang, dark energy, and other unexplained phenomena raise questions about the material view. Science is doing a wonderful job of explaining the mechanics of physical behavior, but it shows little understanding of the myriad mental experiences contained in the vast repositories of literature, art, drama, and other liberal arts fields of study.[26] This is a classic conflict between the cultures of science and the literary arts noted by C.P. Snow.

The mind may be related to matter in profound ways that are beyond our grasp, much as Einstein's relationships between matter, space, and time. Perhaps the uncertainty inherent in quantum physics is attributable to the mysterious powers of human spirit, causing atomic particles to flit between parallel universes. After all, physicists have noted the central role of the observer in shaping physical phenomena.[27] It may turn out that there is something about the processes of life itself that give rise to consciousness: the inner working of cells with a life all their own, the flow of organic materials through a pulsing bloodstream, the intricate chemistry of hormones secreted by a web of glands. A century ago who would have believed that living images of people accompanied by their voices would be transmitted through space and displayed on TV screens? Yet we now know that the universe is alive with *invisible radiomagnetic energy*. One more step in this line of intellectual progress may carry us to a universe pulsing with *spiritual* energy.[28]

There seems little doubt that a realm of human sprit exists, and it seems to operate within a small margin of freedom where life takes place. It's not necessarily "goodness" but includes all aspects of this domain, including destructive beliefs. The only serious question is, where does it come from and how does it work?

I don't think it's possible to know at this time whether science will dispel the soul, or if spirit will gain ascendancy. It is tempting to think these two great trends could converge in time, but for the foreseeable future they seem to be propelling us into the maws of a great conflict now shaping up between science and spirit. Are humans simply biological systems for processing information? Or is there something about life that transcends sheer information? Will the future be dominated by AI systems, the divine spark of life, or a synthesis of the two? Either way, we are in for a fascinating learning experience during the next 20 years or so as our ability to model human intelligence matures. Stay tuned.

MASTERING CONSCIOUSNESS

This is not simply an intellectual issue because the collective awareness of a high-tech, rapidly globalizing world is growing today at lightning speed. Entering the following words in Google in late 2007 produced the following hits:

Mind	421 M	Food	644 M
Spirit	217 M	Sex	459 M
Soul	242 M	Housing	244 M

These are crude indicators, but "mind," "spirit," and "soul" attract roughly the same strong attention devoted to "food," "sex," and "housing."

Culture is a sensitive barometer of change, and it is estimated that one-third of books and movies now reflect spiritual themes. The movie *City of Angels* depicts a world in which spiritual beings like Nicholas Cage watch over our every breath, while in *Ghost*, Patrick Swazye revisits Demi Moore from the afterworld.

The accelerating advances in AI point directly to this mysterious phenomenon we take for granted. Computer power should match that of the brain about 2020, roughly the same time TechCast forecasts AI to be widely used. Honda released the second generation of its android, Asimo, who can run, climb stairs, carry on conversations, escort visitors, and serve coffee. The Japanese and Koreans plan to be selling robots to families about 2010, and our earlier forecast expects they will be commonly used in homes by about 2015. These trends suggest advanced societies will increasingly replace routine human intelligence with reasonably good AI over the next decade.

IBM and others are simulating parts of the brain, and it is conceivable we could produce computer simulations of the entire brain in a decade or two. After all, does it make a difference to replace a neuron with a transistor? The brain contains 100 billion neurons interacting in a network of 100 trillion synapses, which seems impossibly complex. But that's just a technical barrier, like deciphering the three billion bits of information in DNA. If science can break the DNA code, why not the brain? Some seem convinced this means humans will be reduced to mere information systems that are uploaded, stored, and downloaded at will. Computer scientist Marvin Minsky suggests we can eventually solve the population problem by uploading the minds of the world's nine billion people onto a computer that occupies a few cubic meters and costs a few hundred dollars to run.[29]

The rise of AI follows the well-worn historic path in which technology steadily replaced farming and manufacturing, and now promises to automate services and knowledge. But recall that the automation of factories did not produce rampant unemployment as feared. In fact, the reverse occurred as better jobs became plentiful. What we have learned is that automation eliminates routine tasks that can be relegated to machines, freeing people to focus on more creative tasks that can only be handled by humans. I suspect the same will prove true for the automation of thought. Here's how David Brooks at MIT experienced it:

> I quickly established a romantic attachment to my GPS. I could no longer go anywhere without her, and I found comfort in her tranquil and slightly anglophilic voice: "Make a U-turn if possible."

> Since the dawn of humanity, people have had to worry about how to get from here to there. My GPS liberated me from this drudgery. You know how it felt? It felt like Nirvana.

This means the next step in social evolution may be subjectivity. The most powerful, the most intelligent information system may never replace the human spirit, forcing us to recognize that there really is a transcendent dimension governing life. I can vividly imagine complaining as we correct dumb mistakes of robots. We are likely to treat them as backward children, as in Steven Spielberg's movie *AI*.

By clarifying these limits to formal knowledge, the unforeseen consequence may be to magnify our need for all those higher-order qualities we have long resisted. Rather than seek to eliminate these foibles of the human spirit, we might do well to accept the apparent reality that human behavior is intrinsically subjective.

Technologies of Consciousness

One of our greatest challenges is to help people gain control over their increasingly complex lives. Advanced societies face a bewildering maze of information overload, complex relationships around the globe, bureaucratic institutions, and sheer confusion over the pace of change. Stress is the basic cause of 60% of hospital visits.[30] All except saints struggle with some nagging problem that eludes control – weight, smoking, drugs and alcohol, compulsive eating, depression – and the most successful remedies are moral and spiritual, such as the many 12-step programs. Many claim the biggest problems today – crime, sexual promiscuity, conflict, etc. – stem from an irresponsible culture of self-interest, consumerism, and other failures of human spirit.

Spirituality may allow us to feel whole and experience bliss, but its practical function is to alleviate these troubling ills of our time. People need some way to ease their turbulent lives, gain clarity and the moral will for tough decisions, handle uncertainty, maintain cohesive communities, and find some measure of peace and meaning in a changing world. That's why we are increasingly using prayer, meditation, exercise and sports, art and music, ceremonies and rituals, dance, humor, nature, relationships, love, sex, family, community, and a variety of other interventions for mastering inner life. Think "**technologies of consciousness**" – practices that change the state of mind. MIT geneticist Eric Lander said after working with the Dalai Lama: "We should regard [Buddhism] as a refined technology."[31]

I can offer a humble example. My garden never fails to sooth me when I came home stressed, tired, or confused. Over the years I've planted and nourished a never-ending variety of Nature's best. As I watch the garden grow through its intricate life cycles, I am always reminded that these are living creatures in their own right, governed by mysterious forces of nature. The garden is not really mine in that sense, but a gift from whatever creates life and guides it.

In one corner of the garden I was inspired to build a gazebo. The design came to me in a flash. After sketching it out on paper, I built the entire structure myself, marveling as its unusual form took shape. When guests visit, I take them through the garden to the gazebo, and they are always amazed. A calm serenity takes over as the special mood of this place sinks in. I've come to think of it as a "mind machine" because I can count on its ability to create this mood. The following trends highlight the use of such technologies of consciousness in medicine, business, politics, sports, religion, and other arenas:

Technologies of Consciousness

The Mind-Body Connection Evidence on the influence of mental attitudes on our bodies is causing a redefinition of medical care. Dr. Steven Locke at Beth Israel Hospital called it a "revolution" akin to the advent of penicillin, while Dr. Sandra Levy at the Pittsburgh Cancer Institute claims the evidence is now "indisputable" and "could change the entire face of medicine." Most medical schools include acupuncture, massage therapy, meditation, mindfulness, and other alternative practices in their curriculum. Half of Americans use these practices, and many have been verified to be effective.[32]

Spirituality is Productive Spiritual practices are entering the workplace because they produce results. Lawrence Perlman, CEO of Ceridian, said, "Ultimately, the combination of head and heart is a competitive advantage." Executives at Aetna, Monsanto, McKinsey, Medtronic, AOL, Raytheon, and Silicon Graphics meditate daily. At Lotus, a "Soul Committee" was established to shape values. One organization starts meetings with a minute of silence, and I know an executive who lights a candle when a visitor comes to talk. A "spiritual audit" of 200 corporate leaders found "spirituality is one of the most important determinants of performance."[33]

The Boom in Yoga Yoga has grown into a major industry. During the past few years, the leading magazine *Yoga Journal* has spurted to a half million readers as 15–20 million Americans have learned to take up the discipline. The number of instructors has blossomed from 2,000 five years ago to 14,000 in 2006.

Spirit and Politics Spirit is involved in politics because beliefs are a source of power. This was illustrated when politicians attended a rally held by evangelist Billy Graham. One said, "We think of ourselves as being in the business of motivating people. But that! That was power." Presidential campaigns are increasingly marked by pledges to support family values, moral conduct, and religious beliefs, both from the right and the left.

Playing the Mind Game There is a growing realization that strength, timing, and other athletic skills emanate from the mind. Coach Phil Jackson of the Los Angeles Lakers helps players achieve their best by reading Nietzsche, and Brenda Bredemeir at Notre Dame calls her approach "holistic" and "spiritual." Many athletes practice by visualizing perfect executions. Tiger Woods forms a mental image of each stroke being played out.

The New Religiosity Religious dogma is yielding to personal beliefs and practices drawn from a variety of faiths, often including care for family, community, and the environment. Richard Cimino and Don Lattin have studied this new form of religiosity, and they conclude: "Spirituality and religious faith are increasingly viewed as private matters."

Drugs and Spirit The war on drugs seems futile because people find benefits in medicating the mind. We have gone from marijuana and cocaine to some 200 recreational drugs and more are being invented. An experiment was conducted at Johns Hopkins University in which 36 mid-aged volunteers were given psilocybin (mushrooms) while a control group was given a placebo. Those taking the drug called it the most significant experience of their lives. In a follow-up two months later, 79% reported increased well-being, which was confirmed by family and friends.[34]

Sex as Spirituality? The popularity of birth control, Viagra, and online porn make it clear we've come a long way from the Victorian days of sexual repression. Freud would be amazed to see the way Madonna, Britney, and other celebrities celebrate sex today. In a recent survey, 55% of Americans said sexuality is "an integral part of their spiritual life." Most called it "A gift from God."[35]

A Global Movement Businessman and author Paul Hawken finds there are one to two million groups around the globe dedicated to solving the dilemmas of our time, mobilizing many millions of people. He sums it up this way: "The promise of this unnamed movement is to offer solutions to what appear to be insoluble dilemmas:

poverty, global climate change, terrorism, ecological degradation, polarization of income, loss of culture."[36]

Forecast: An Age of Consciousness about 2020–3030

This is hardly a complete summary, and there are counter-trends to be sure, such as the rise of atheism. But this chapter illustrates that powerful subterranean forces are relentlessly moving society toward a global form of consciousness simply out of the need to withstand a more difficult world. Often this takes the form of conflicts that are crying out for resolution.

Here's a quick summary of prominent issues. As globalization dramatically increases the load on the environment, people are being mobilized toward creating an ecologically safe global order. Relentless technological advances are forcing us to use wisely the power of AI, biogenetics, and nanotech. Young people in almost all nations are today connected through digital media into a common global culture. The clash between Islamic zealots and the West reminds us of the brutal need to form a workable system out of hundreds of diverse cultures. Conervatives and liberals in the U.S. rage against each other, and there is an escalating conflict between the rich and the poor.

The great challenge facing civilization is to bring these islands of consciousness under control, much as we tamed nature with farming, created material abundance, formed global institutions, and are now harnessing the power of information. The transition to a global order of nine billion people demanding modern lifestyles will require a fundamental change in consciousness – or the collapse of an ecosystem, climate change, nuclear war between nations, and other megadisasters will serve to prod us along. Regardless of whether we think of this in biological, digital, or spiritual terms, it would be a good idea to understand this growing power of consciousness and learn to guide it.

I estimate an "Age of Consciousness" is likely to occur sometime between 2020 to 2030. It may be called a "Crisis of Maturity," a "Global Community," or an "Age of Wisdom," but I think it's coming. The economic trends are compelling. By 2030, the *Economist* magazine forecasts that China, India, Russia, and other large developing nations will approach economic maturity. China's economy will match that of the U.S., and the number of people living at industrial standards will reach five billion out of a global population of about eight billion. Most importantly, world GDP will rise from its present level of roughly $50 trillion to $185 trillion, almost a four-fold increase, with concomitant rises in environmental load, demands for energy, intensifying market competition, and other challenges.[37]

It is almost inconceivable that the world can survive these challenges with our present attitudes, so a transition to global consciousness seems almost inevitable. The Dalai Lama put it best: "The root cause of these manmade problems is the inability of humans to control their agitated minds.[38] Please note that I use the term "crisis of maturity" in the Asian sense, meaning a critical point of transition, rather than the Western sense of a catastrophe. This crisis could possibly be headed off even now if world leaders are able to mobilize attention. Given the enormous difficulty of bringing about such massive changes, it seems likely that action will be delayed until it is unavoidable at about 2030, as the scenarios in the next chapter illustrate.

I fully realize the idea of an age of consciousness may sound naïve, with violence and selfishness raging throughout the planet. Obviously, people will not suddenly become altruistic. That's not the point. The point is that we are beginning to recognize that all those complex, emotional, conflict-ridden aspects of human behavior may represent attempts to cope with the spiritual dilemmas of life, often in crude ways. As AI automates much of the routine mental tasks that now occupy us, I think we are likely to realize that we live in what is basically a spiritual world guided by competing visions, values, and beliefs, with its good, bad, and indifferent forces. We may then learn to cultivate this new terrain seriously, rather than muddling through with ignorance, dogma, denial, and superstition.

Most scientists today challenge this spiritual view as wishful thinking, a misguided need to believe there is something unique about humans, whereas they believe human thought is reducible to sophisticated AI. Powerful forms of AI and all the other technologies noted in this book will certainly prove essential. But to believe that intelligent machines alone will resolve the obstacles now creating a global crisis of maturity – that would truly be wishful thinking.

This approaching transition will hardly be Utopian because the power of spirituality is easily misused, like anything else. Religious and political zealots, for instances, have always forced civilization through endless wars and ideological conflict. It may even be that the future will pit more intense moral differences against one another, creating biblical-like battles between good and evil.

Spirit does not necessarily imply "goodness," but a higher state of mind that can take almost any form. The great challenge now facing civilization is to accept this mysterious power and use it to guide our complex lives more carefully and with greater meaning. This humble but all-important task, I submit, will constitute the next great frontier beyond the Information Age. The Information Age provides unbounded knowledge, but the Age of Consciousness promises values, wisdom, meaning, purpose, beliefs, vision, and other intangibles we use to organize knowledge.

All of the nagging problems that seem insolvable today basically flow out of today's state of consciousness, so the basic solution is to alter consciousness. This sounds idealistic, but that's what we said about knowledge not too many years ago. Who would have believed we would spend half the day staring into a PC?

Notes

1 The author acknowledges gratefully the constructive guidance of Thomas Lombardo, Professor of Psychology at Rio Salado College, for helping with technical issues in this chapter.

2 Quoted from *The Sun* (November 1989); "A Memoir of War," *Washington Post* (July 13, 1986); "Right At Home," *Washington Post* (November 14, 1998).

3 The basic function of living creatures is to organize the brain into a coherent system for focused understanding. Bruno von Swindereren at the Neurosciences Institute in San Diego, says, "Attention is a whole brain phenomenon, a binding together of different regions." And Simon Conway, professor of paleobiology at Cambridge, called it "An antenna." Adam Zeman says, "Consciousness bridges perception and action, the events we perceive and the ones we bring about." Adam Zeman, *Consciousness: A Users Guide* (New Haven: Yale Press, 2003); Douglas Fox, "Consciousness in … a Cockroach?," *Discover* (Top Ten Stories of 2006); "Do Fruit Flies Dream of Bananas?," *New Scientist* (February 14, 2006).

4 *Decisions, Uncertainty, and the Brain: The Science of Neuroeconomics*, Paul Glincher (MIT Press, 2003).

5 David Myers, *Intuition: Its Powers and Perils* (Yale, 2003); Antonio Damasio, *Looking for Spinoza: Joy, Sorrow, and the Feeling* (NY: Harcourt, 2003); Edward O. Wilson, *Sociobiology: The New Synthesis* (Belknap Press, 2000); Steven Pinker, *The Blank Slate: The Modern Denial of Human Nature* (NY: Viking, 2002); Carey Goldberg, "A Question of Will," *Boston Globe* (10/15/02); "In Search of the Buy Button," *Forbes.com* (September 1, 2003); John Horgan, "The God Experiments," *Discover* (January 2006).

6 Steven Pinker, a psychologist at MIT, concludes that human behavior "will increasingly be explained by neuroscience, genetic inheritance, and evolution…The soul is the information processing of the brain." See Pinker, *The Blank Slate, Op. Cit.*

7 Antonio Damasio, *Looking for Spinoza, Op. Cit.*

8 "Gelertner, Kurzweil Debate Machine Consciousness," Kurzweilai.net (12/11/2006).

9 "On the Luminosity of Being," *New Scientist* (May 24, 2003).

10 Shankar Vedantam, "Tracing the Synapses of Spirituality," *Washington Post* (June 17, 2001).

11 See Willis Harman, *Global Mind Change* (Indianapolis, IN: Knowledge Systems, 1988).

12 "The New Science of Mind," *NewsWeek* (September 27, 2004); Rob Stein, "Researchers Look at Prayer and Healing," *Washington Post* (March 24, 2006); "Saul Bellow," *The Week* (April 22, 2005); Steven Kotler, "Extreme States," *Discover* (July 2006); "On the Edge of the Known World," *NewScientist* (March 13, 2004); Shankar Vedantan, "Near Proof of Near Death," *Washington Post* (December 17, 2001); Richard Morin, "Calling Dr. God," *Washington Post* (July 18, 2004); "The Nobel Effect," *Newsweek* (January 29, 2007); "Rupert Sheldrake," *Seven Experiments that Could Change the World* (NY: Riverhead Books, 1995); Jeff Levin *God, Faith, and Health* (NY: Wiley, 2001). U.S. and Russia devoted two decades to "remotely viewing" enemy activities, often with great success. Warrant Officer

Joe McMoneagle, one of the best psychics in the U.S. program, was first to identify the existence of Russia's Typhoon submarines.

13 Shankar Vedantam, "A New Thinking Emerges About Consciousness," *Washington Post* (May 20, 2002).

14 Willis Harman did a wonderful job of explaining this perspective decades ago and anticipating the global shift of consciousness. See Harman, *Global Mind Change*, *Op. Cit.*

15 A good example of serious attempts to define the laws of spirituality is represented by the work of Joseph Campbell, who found common themes in spiritual experience. See Campbell, *Myths to Live By* (NY: Viking Press, 1972).

16 Edge.org (November 20, 2006).

17 "90% Believe in God," *Newsweek* (March 31, 2007); "Americans May be More Religious," *Washington Post* (September 12, 2006); "Religion, the Eternal Growth Industry," *Washington Post* (January 4, 2004).

18 Kevin Sullivan, "Novel Faiths Find Followers in Russia, *Washington Post* (July 17, 2007).

19 The near universal need for faith of some type is well explained by Mark Lilla, *The Stillborn God* (NY: Knopf, 2007).

20 Sam Harris, *End of Faith* (NY: W.W. Norton, 2004) and Richard Dawkins, *The God Delusion* (U.K.: Bantam, 2006).

21 Reported in www.Edge.com (November 20, 2006) and Marc Kaufman, "A Meeting of Minds," *Washington Post* (September 21, 2003).

22 Halal, "The Life Cycle of Evolution," *Futures Research Quarterly* (2004) 9:1.

23 Sperry, R.W., "The Impact and Promise of the Cognitive Revolution," in R.W. Solso and D.W. Massaro (eds), *The Science of the Mind: 2001 and Beyond* (New York: Oxford University Press, 1995).

24 Sharon Begley, *Train Your Mind, Change Your Brain* (NY: Ballantine Books, 2007) and Norman Doidge, *The Brain that Changes Itself* (NY: Viking, 2007).

25 Ilchi Lee, *Human Technology* (Sedona Arizona: Healing Society Press, 2005).

26 C.P. Snow described this isolation of the two cultures of science and art.

27 A good summary of the "consciousness creates reality" view is provided by Wikipedia in "Consciousness Causes Collapse."

28 The view of a spiritual universe is well described by Teilhard de Chardin, a Catholic priest and anthropologist. See *The Phenomenon of Man* (NY: Harper Collins, 1975).

29 Ronald Bailey, "Would you give up your immortality to ensure the success of a post-human world?" www.reason.com (July 27, 2007).

30 Michelle Conlin, "Meditation," *Business Week* (August 30, 2004).

31 *Washington Post* (2003).

32 "The New Science of Mind & Body," *Newsweek* (November 27, 2004).

33 Peter Pruzan and Kirsten Pruzan Mikkelsen, *Leading with Wisdom: Spiritual-Based Leadership in Business* (Sheffield, U.K.: Greenleaf Publishing, 2007); Michelle Conlin, "Religion in the Workplace," *Business Week* (November 1, 1999); "The Jack and Herb Show," *Fortune* (January 11, 1998); "Human Potential," *Executive Update* (May 2000).

34 "Hallucinogens Induce Lasting High," *The Economist* (July 13, 2006).

35 "God's Gift," *News Week* (October 2, 2006).

36 Paul Hawken, "Something Earth-Changing is Afoot," *Orion Magazine* (May–June 2007).

37 Robin Bew, Chief Economist, *Focusing on the Horizon* (*Economist Intelligence* Unit, 2005), and *Long-Term Forecasts*, www.Economist.com.

38 "Brute force can never subdue the basic human desires," *Washington Post* (October 21, 2007).

Chapter 10

Scenarios: A Virtual Trip Through Time

I find it fascinating and humbling to witness the harsh lives people struggled through when the world was very different. One of my favorite movies is *Quest for Fire*, reliving the ordeals prehistoric men faced in keeping the hearth burning, fighting ferocious beasts, and surviving massive wounds without medical care. The medieval period is equally intriguing, with its feudal lords and bishops ruling ignorant peasants. Even the last few decades were replete with ideas soon to be overthrown, like the simmering conflict between Capitalism and Communism and the innocence of communications before the Internet.

These remarkable changes in the basic conditions of life remind us that civilization is passing through a series of stages that supersede one another. Much of what was obvious to the ancient Romans – slavery, killing as sport, a panoply of Gods – looks pitifully brutal and naïve today. In 16th century Paris, the royalty and their friends would delight themselves by burning cats alive. It's tempting to feel smug at the distance we have come since then, but it is almost certain that life today will appear equally brutal and naïve as the Technology Revolution runs its course.

This chapter examines all of our forecasts longitudinally in an attempt to grasp how the conditions of life are likely to evolve over the next few decades (Figure 10.1). Think of it as "macro-forecasting." Ideally, I'd like to penetrate the shell of technology, economics, and institutions to experience the look and feel of how we will live in 20–30 years or so when most of the world is industrialized and educated, when artificial intelligence (AI) is everywhere, and a working global community emerges to govern the planet.

Anticipating the future is more fascinating than marveling at the past. The changes are arriving far more quickly, they should prove more dramatic, and the challenges will be greater than anything we have met. Wouldn't it be useful to understand how the future world will work? Explore the new joys to be discovered and the pains we must bear? Ponder what would be required to realize these possibilities?

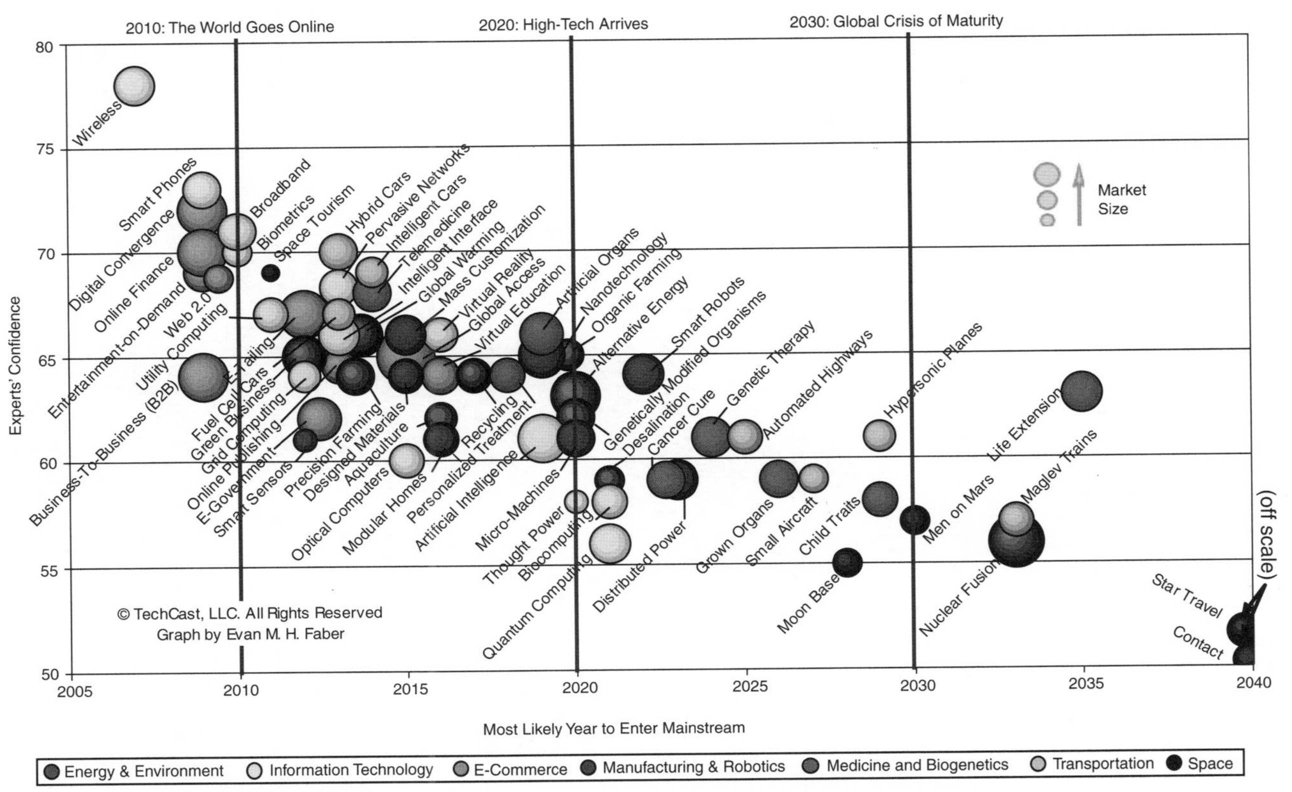

Figure 10.1 Longitudinal Summary of Forecasts

Join me on this virtual trip through time. We'll take it decade by decade, combining breakthroughs in all fields to form a realistic portrayal of how life is likely to evolve. We don't hope to get the details right, of course, or how specific events will unfold. And there is a margin of uncertainty surrounding each forecast of about +/– 2 to 5 years. But the forecasts covered here lay a pretty solid foundation for the emerging social order. To enhance our understanding, I cluster related technologies into waves of innovation that put a distinctive stamp on each decade. My goal is to identify the dominant themes of each period, provide a few illustrative examples, and assess the social impact. The following scenarios outline the most probable perspective of how history will be written looking backward from the year 2050.

Summary of Scenarios

2010: The World Online This decade should continue to focus on intelligent advances in information systems and e-commerce. The world in 2010 is almost certain to be smarter, faster, and fully wired, setting the stage for the breakthroughs to come.

2020: High-Tech Arrives This decisive period should see major breakthroughs in high-tech. Green business, alternative energy, and other ecological practices are likely to ensure ecological sustainability. AI should permeate life, and the next generation of quantum/optical computing will permit huge advances in telemedicine, virtual education, and e-government. Biotech should mature, providing personalized medicine, genetic therapy, cancer cures, and other advanced healthcare.

2030: Crisis of Maturity Industrialization will reach most developing nations at this point, with as many as five billion people living at modern levels of consumption. Although technological powers will be vast, intercultural conflict, weapons of mass destruction (WMD), and threats of environmental collapse are likely to grow into such challenges that they force the move to a global shift-in consciousness.

2040–50: Global Order Civilization has withstood the Fall of Rome, World Wars I and II, and threats of nuclear holocaust, and it will probably survive globalization. The challenges facing civilization are likely to be resolved to form a modernized, fairly harmonious globe, somewhat like a far larger and more diverse version of the U.S. or E.U. Local wars, ecological disasters, and other mishaps will continue, of course, but limited to the normal dysfunctions of any social system.

SCENARIO 2010 +2/−1 years

The World Online

In retrospect, it's now clear that the period of 2010 proved seminal as people around the globe came online with full communications capabilities for the first time. As wireless increasingly connected almost anything, digital convergence spread information throughout the system, while higher transmission rates moved it at lightning speeds. At about 2010 there was a palpable sense that the entire globe had become alive in an organic web of smart phones, iPods, laptops, PCs, TVs, and various other devices filling the air with information, knowledge, and intelligence.

As Internet traffic rose steeply, "computer utilities" provided cheap computer power and software applications on demand while huge grid networks served special industries, regions, and projects. A ten-fold gain in broadband speeds carried the heavy load of communications, and biometrics assured security. Huge improvements were made in speech-recognition, permitting the bulk of common transactions to be handled by simply speaking clearly to machines. Web 2.0 moved into the mainstream, creating participative forms of e-commerce, media, etc. Corporations learned to tailor products and services to the constantly changing tastes of diverse customers, and governments built online systems that served the citizenry wherever they were.

Somewhat later, a profusion of teleliving erupted as full-screen video communications with intelligent voice controls soon appeared in business, government, law, and most other fields. More convenient online selling, shopping, entertainment, and finances restructured the economy into a seamless network of instant control. Religious services became wonders of video presentation, with active involvement by worshipers online.

In 2015, Microsoft saw the obsolescence of windows as the keyboard and mouse gave way to an intelligent interface. The company's long dominance of the computing industry came to a close when a flurry of competitors introduced software that allowed intelligent avatars to handle routine tasks on demand. Youngsters took delight in giving their virtual assistants clever names like "dog," "honey," and "Harriet," and sent them off to deal with the virtual assistants of their friends, who had similar names.

In a more practical vein, developing countries came alive as cell phones and $100 laptops integrated farmers, businessmen, and educators into the global economy. Automatic language translation was a big aid in communicating across cultures, but the real push came from digital media that made films, music, and TV a common form of entertainment shared by almost anyone.

SCENARIO 2020 +/– 3 years

High-Tech Arrives

Three major developments were launched during this period. The first was the transition to sustainability, initiated when corporate management moved *en masse* to ecologically safe business practices in 2015. As the first wave of ventures into biofuels, advanced solar cells, fuel cells, hydrogen extraction, and ecological management became successful, it suddenly seemed as though a switch had been flipped when attitudes to the environment turned suddenly from caution to enthusiasm.

Hybrid cars became common and the first commercial fuel-cell autos were released by GM to great fanfare. With less expensive photovoltaic cells using nanotech, panels began appearing on car roofs and trunks, and light-weight auto bodies made of carbon-reinforced plastic were introduced. Average fuel efficiency rose to almost 80 mpg. As acceptance of these innovations spread, political pressures caused most nations to pass laws that either taxed carbon or used a cap-and-trade scheme to discourage the release of greenhouse gasses, taking a major step to reduce global warming. The rush to find business opportunities in this transition produced intense competition in precision-farming, recycling, organic farming, Genetically Modified Organisms (GMO), desalination, and distributed power, with speculators bidding up the stocks of new companies to illogical heights, reminiscent of the dot-com boom. The transition to sustainability passed a critical point in 2020 when 30% of all energy came from alternative fuels.

The second major breakthrough was the transition to the next generation of IT and the maturing of AI. IT moved into high-gear in 2015 when the first optical computers calculating with light signals vastly increased the speed of computation. This was followed by the introduction of commercial quantum computers in 2021, along with breakthroughs in biocomputing and nanocomputing.

With all this computational capacity, AI was able to perform most human tasks reasonably well, replacing the routine work of teachers, physicians, lawyers, and most other professions, who could now focus on giving clients personal attention, thinking strategically, and conducting research. Thoughts of computers passing the Turing test remained distant, however, as any resemblance to actual humans seemed remote, with people constantly complaining about "dumb machines" and "stupid mistakes." Consolation was at hand, however, as Virtual Reality (VR) came of age, with people enthralled by virtual visits to outer planets, beneath the oceans, and – you guessed it – virtual sex. Just as the porn industry was first to use the Internet effectively, online sex thrived using high-definition, interactive virtual reality. For a small fee you can now enjoy intimate moments with lifelike models of your favorite stars online.

The third breakthrough was the fulfillment of DNA's potential to allow mastery of biogenetic medicine. Later in the decade, personalized treatments became widespread, cancer was brought under control using nanobots and advanced drugs, genetic therapy was able to correct many portions of faulty DNA, and genetically identical organs could be grown to replace failed body parts. Using the new generation computers, micromachines, and designed materials, artificial organs were able to substitute for lost limbs and failing hearts, livers, and kidneys.

These three waves of progress set the tone for high-tech life in modern nations. Spacecruisers were routinely ferrying tourists into Earth orbit and around the Moon, creative home designs proliferated as modular housing spread in popularity and convenience, and almost anything could be ordered online cheaply customized to personal taste. Homes, offices, cars, healthcare, and the natural world became a dense web of networks interacting through embedded sensors. Electrical power became cheap and reliable as power grids were decentralized into networks of windmills, fuel cells, solar banks, and other alternative energy systems. But the old "blackouts" gave way to "whiteouts" in which the IT networks crashed, leaving people stranded in well-lit homes.

SCENARIO 2030 +/– 5 years

Crisis of Maturity

Riding on the breakthroughs of 2020, this decade soared as worldwide innovation and productivity spread economic growth to almost five billion industrialized people around the globe. And with the transition to ecological sustainability underway, the power of AI automating huge swaths of work, and people enjoying the benefits of genetic medicine and the adventure of space travel, prospects seemed infinite.

Major highways were automated in 2025, zipping people along congested roads at 70 mph with a big drop in accidents. A lush variety of private air travel sprang up in major cities as executives and professionals bought planes fitted out as flying versions of the family van. Hypersonic flights carried people from New York to Beijing in three hours, and Maglev trains speeded pampered passengers between urban hubs at 400 mph. Space became a major industry when a permanent colony was established on the Moon in 2028, soon followed by a manned landing on Mars.

Optimism peaked in 2030 when the world's first fusion nuclear reactor went online and began delivering the power of the sun to homes and offices. Average life spans approached 100 years. Terrorism, pockets of poverty, and economic failures persisted, but these dilemmas were thought to be unsolvable while the powers of human ingenuity promised a boundless future, sending the Dow Jones Industrial Average to a record of 100,000.

The party crashed when tensions between rival Mid-Eastern nations erupted. Efforts of the international diplomatic community were to no avail, and the first nuclear exchange since Hiroshima and Nagasaki occurred in 2031, wiping out Tehran, Karachi, and Tel Aviv, killing tens of millions and making much of the Middle East uninhabitable. A stunned world reacted swiftly but without conviction. Stock markets plunged around the globe, economies swooned, and the planet was gripped with a fear not known since the bombings of London in World War II.

The president of the United States, a charismatic woman, called her counterpart in China, and together the two superpowers scheduled an emergency meeting of the UN Security Council. After the usual procrastination, the urgency of the crisis forced them to consider seriously how they could convert this calamity into an historic opportunity to rid the world of WMD. The secretary general of the UN was an Indian who loathed war, and together they marshaled suport for a resolution banning WMD. Resistance seemed overwhelming and talks went on endlessly, until the two presidents and the UN Secretary made a trip to the warring nations. Making personal pleas, they persuaded the combatants to renounce violence if other nations disarmed and the UN, the U.S., and China served as mediators to forge a peaceful settlement. It seemed the only reasonable thing to do.

The news brought a sea change in international relations. The hard work of producing official agreements was left to diplomats and bureaucrats, while the recently appointed Catholic Pope called on religious leaders to attend a conference to reconcile antagonism between Islam, Christianity, and other religions. The debate stalled repeatedly until Tenzin Gyatso, the aging Dalai Lama, made a rare appearance to plead for his visionary ideal of non-violence, all faiths accepting the basic unity of humanity, and a commitment to preserving the planet as their sacred common home. The Dalai Lama was revered, and his influence helped them to draft a set of common principles that were supported by all religions and even atheists.

The conference was telecast around the globe, demanding that the long history of warfare finally be brought to a close. Buoyed by this unprecedented success, agreements were also reached on a host of other long overdue resolutions, including one assuring nations of UN assistance in economic development and open-markets. The UN Secretary General called it "The start of a true global civilization."

The U.S. President seized the moment to call for expanding the role of global corporations into a broader working community of employees, clients, and the public in addition to the rights of capital. American business had been moving in this direction since the Green Business Movement of 2010 and the revolt of the middle class against the extravagant pay of CEOs, athletes and movie stars. It also made sense to the new Gen Y breed of corporate executives.

The President saw the political potential of claiming this advance as a logical extension of America's Democratic ideals, so she brokered an agreement with the U.S. Chamber of Commerce and other prestigious business organizations. Together they proposed the voluntary adoption of "Democratic Enterprise" to make corporations more productive partners with people around the world. Most large corporations welcomed the news as they had been pursuing variations of this basic idea, and it helped now that the concept was defined and widely accepted.

These acts of leadership calmed anxieties, and encouraged a more deeply participative world that better harnessed the talents of ordinary people. Business was admired for its goals of working with its "social partners" to improve their common welfare, while disadvantaged people thrived now that work became a network of small self-managed enterprises they could control themselves. The excessive consumption of "The Second Gilded Age" inaugurated by President Reagan diminished as old icons fell. Don Trump, now an eccentric, became a pariah as great wealth lost its glamour and business no longer focused entirely on money. American politics was forced to stop accepting campaign contributions as people admitted they were corrupting, and the nation moved to publicly financed elections. You could hear many people saying "It's great to feel proud of being an American again," and the President of the U.S. said with her characteristic humility, "It's about time."

SCENARIO 2050 +/– 10 years

A Global Order Emerges

After the decisive changes of 2030, the world groped its way toward a mature state of development about 2050 marked by a few distinctive features. Global population peaked at nine billion people with seven billion living in knowledge societies and the rest remaining at industrial or agrarian stages. Worldwide information grids are used to manage the complexity of this huge global system, and the bulk of people work in various fields that push technological breakthroughs. The seven-fold growth of industrialization over the past few decades forced techno-economic systems to be made ecologically benign, yet minor environmental crashes still occur.

Most nations are part of a functioning global community, but decentralized governance also nourishes wide diversity in traditional cultures. These differences are held together by an international culture that agrees on common standards of behavior and general set of values that unify various traditions into a functioning whole. Society works hard to curb the destructive tendencies of people with limited success, so occasional wars, crime, and other forms of violence persist.

Terrorism receded as the cultural gap between Islam and the West diminished with these changes. Politicians from both sides reached agreements to have their nations work together, but the real impetus came from the new democratic corporations that sought out poor people for growth opportunities. The Middle East realized its potential as its entrepreneurial talents were put to use in developing their nations.

Now that the planet contains billions of educated people organized around IT networks that allow close collaboration, this vast global brain is being harnessed to discover the deeper laws of physics, which may allow humans to explore the universe. Fundamental breakthroughs in the workings of space-time-matter led to today's "Star Finder" probes that use teleportation technologies to contact alien planets. Initial results look promising, and if we can gain a little more confidence in this system, we should soon be able to launch our first manned flight to another star system.

UNDERSTANDING OUR PLACE IN HISTORY

Obviously, things are not likely to work out as neatly as they are outlined here, but that's beside the point. This mental exercise of virtual time travel through progressive scenarios is not intended to get the details right. Sketching out a compelling story of how technology is moving to higher stages of development helps us grasp the trajectory of history. The specific facts can't be known, but the broad arc of this path to a unified world is clearly marked by the evidence in this book. I'd like to draw on previous work to put this in perspective as the "Life Cycle of Evolution" (LCE).[1]

The LCE comprises that long-term trend spanning the rise of biological species, nomadic humans, Agrarian Civilization, the Industrial Era, a Service Economy, today's Information Age, and possibly an Age of Consciousness. This larger sequence of developmental stages reveals how technological change drives a cycle of organic development for the entire planet, although on a scale of such enormous magnitude that almost defies comprehension. We are not used to thinking in such daunting terms, but the Earth appears to be evolving through its own life cycle, roughly similar to the life cycle of ordinary organisms.

Table 10.1 summarizes the seven stages that are believed to comprise the entire LCE using data from well-established sources.[2] Note that all stages are driven by technological progress, including "soft" forms of technology. For instance, biological evolution has been called "the technology of life." In a technical sense, species evolve when information stored in two sets of DNA molecules is combined sexually to produce better adapted offspring. Even spiritual practices are used to alter awareness, which we called technologies of consciousness. These different forms of technology all evolve through the same Darwinian

Table 10.1
Technology's Role in Evolution

Stage of Evolution	**1 Biological Era**	**2 Nomadic Era**	**3 Agrarian Era**	**4 Industrial Era**	**5 Service Era**	**6 Knowledge Era**	**7 Global Era**
Dominant Technology	Genetics	Primitive Tools	Farming	Physical Technology	Social Technology	Information Technology	Technologies of Consciousness
Beginning of Era	Creation of Life 4 billion BC	Development of Humans 3 million BC	Agrarian Revolution 7000 BC	Industrial Revolution 1850 AD	Post-Industrial Revolution 1950 AD	Knowledge Revolution 2000 AD	Crisis of Maturity 2030 AD + 10/–5 yrs
Historic Outcome	Species	Clans & Tribes	Civilization	Mass Production	Large Organizations	Information Networks	Global Order

process as life experiments through competition for the survival of the most fit; the only difference is that the struggle takes place among competing technological artifacts rather than biological organisms.

Each stage of social development uses its unique technical base as the foundation to construct a different social order. The prevailing technology determines the dominant type of work people do, their social institutions, and central values – in short, the nature of the entire society. Marshall McLuhan noted, "Any technology creates a totally new human environment," and Arthur C. Clarke summed it up as simply, "Tools invented man." Technology is derived from creative thought, cultural conditions, visionary people, and a host of other sources, but conditions do not change until the power of innovation is applied widely to alter the social order. In short, technology is the fulcrum on which cultural forces leverage change throughout society.

The long-term trend formed by the data in Table 10.1 is plotted in Figure 10.2 to illustrate the complete LCE. The time horizon is portrayed in logarithmic scales to compress these enormous time differences into a comprehensible figure.[3] On an ordinary scale, the curve would simply run flat and then turn up at a right angle. This shows that the entire rise of civilization takes place in an infinitesimally short

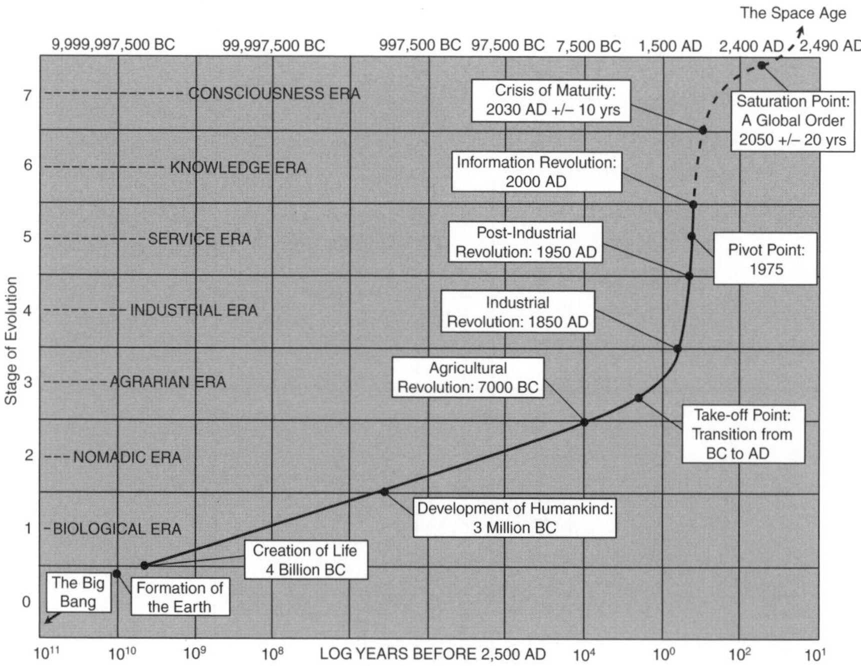

Figure 10.2 The Life Cycle of Evolution

period by cosmic standards, as though the planet suddenly came alive in a flash of creative energy.

Spread out on a logarithmic scale, however, we see a well-ordered progression of developmental stages making up the life cycle of the planet. This is the same S-curve that characterizes the growth cycle of all life forms: a culture of bacteria, a human being, or the life of a planet. In other words, our limited perspective obscures this realization that the entire Earth constitutes a single growing organism in its own right. James Lovelock first proposed this concept which he called Gaia in 1970.

Progress along the LCE has accelerated rapidly by almost any measure of change: world population, the speed of travel, energy usage, and countless other indicators all bend sharply upward. This pattern of accelerating growth becomes apparent by noting that the times between stages are consistently shorter by orders of magnitude: 4 BILLION years of biological evolution were required to develop humankind; 3 MILLION years passed before the onset of civilization; 9 THOUSAND years later industrialization occurred; 1 HUNDRED years after that the Service Era began; and 5 DECADES were needed to reach a Knowledge Age. Our forecasts and this trajectory suggest the LCE should culminate about 2050 in a functioning, unified world. Cosmologists define it as a "Type I Civilization" able to integrate its efforts to manage the planet.[4]

This broader perspective has emerged only recently as IT began to clarify our view of the evolutionary pathway. Now the Earth seems alive, growing like a huge organism through a life cycle of development that may occur on other planets throughout the universe. Biological species are part of this organic process of planetary development, but they constitute the minutia of evolution, rather like the countless subsystems that constitute the body of any larger organism. Trying to understand evolution by focusing on individual species is akin to studying human behavior by focusing on the cells of the body.

We have only one planet to examine as yet, but despite wide cultural diversity around the globe, civilization everywhere has traced roughly the same path marked by these seven stages. Science is pursuing the search for extraterrestrial life because it is estimated that living planets must number in the billions. When and if life is discovered elsewhere, it will be interesting to see if similar stages have been followed there as well.

Please do not think this course of development promises a Utopia. On the contrary, the path outlined here is exceedingly demanding. Progress struggles up a series of stages that produce both great costs as well as great gains. Modern life is so vastly improved over the harsh conditions of ancient societies that average lives today would have been considered opulent in then past. But a great price has been paid in congestion, pollution, bureaucracy, and other problems that did not exist before. And although the Information Age may alleviate these ills, it in turn produces information overload, exploding complexity, globalization, WMD, and other new disorders.

Evolution seems to bring mixed blessings, then, because each stage solves the challenges of the preceding era – but it introduces new challenges that are even tougher. As advanced technologies increase human capabilities, they also produce greater dangers endemic to a more difficult world. The result is that civilization must continually struggle forward to contain the turbulence produced by its own progress.

This continual trading of old problems for new problems illustrates that social progress is neither "good" nor "bad" in the ordinary meaning of these terms. Progress magnifies the existential dilemmas of life, and so its most outstanding feature is that life at advanced stages becomes more *intense*. We increasingly require the careful use of more sophisticated technologies, involving greater complexity and risk, to carry out grave new responsibilities, with more profound consequences. The dilemmas that must be surmounted may be painful, but they appear to be an intrinsic part of this process because they transform us into more fitting vessels of the spiritual energy drawing life on. The philosopher Nietzsche put it well: "The universe is a machine for making Gods."

The great significance of this event can be best grasped by comparing the LCE to a human life cycle. People pass through similar stages, and today the world seems to be working out roughly the same crisis that tests every youth. The typical adolescent has reached almost complete physical growth but has not yet developed the social, intellectual, and moral skills needed to function in a complex world. This is almost precisely the state of modern nations. Advanced societies today possess physical technologies that can destroy all life, but they have not learned to control this power with collaborative institutions, useful knowledge, and shared values.

The LCE serves a useful analogy, because it is roughly similar to the challenges people wrestle with in their daily lives. The main difference is that the planet's life span covers eons of time, progress is imperceptibly slow, the system is infinitely larger and more complex, and the stakes are vastly higher. From this view, the crisis of global consciousness in Scenario 2030 represents a rite of passage to maturity for humankind, just as the crisis of puberty ushers the child into adulthood.

Passing successfully through this crisis of maturity is not inevitable, of course, but it is most likely because this is a natural process that far exceeds human powers. If the planet is a great organism moving through its life cycle, our crises are mere blips on this grand process of global development. The LCE shows that civilization has evolved steadily from tribes, to cities, to nation states, to superpowers – leading toward the next logical step of a unified world order. Some may be skeptical about this prospect, and others may feel joyous or even fearful. Regardless of which reaction turns out to most justified, I marvel at what a great privilege it is to witness this maturing of civilization's long journey.

Notes

1 Halal, "The Life Cycle of Evolution," *Futures Research Quarterly* (2004) Vol. 9, No. 1.
2 Carl Sagan, *The Dragons of Eden* (NY: Ballantine Books, 1977).
3 The horizontal dimension is a 10 cycle logarithmic scale needed to display the time differences between various stages of evolution. Because log scales cannot have a zero point, all times must be expressed along a single direction; the scale was therefore defined as "Years Before 2,500 Years AD" and because this reference point spreads the present stages of evolution out conveniently. The vertical dimension is an ordinal scale corresponding with the rank order of the stages shown in Table 10.1.
4 Michio Kaku, *Visions* (NY: Anchor Books, 1997).

APPENDIX

The TechCast Expert Panel

TechCast is a virtual think tank in the sense that we integrate the knowledge of roughly 100 experts working online around the world. To maintain excellence, we require applicants to have advanced degrees, experience in their field, patents and publications, and other proven results. Some are invited on the basis of their reputations, and others provide their qualifications on the site and pass a multiple choice quiz. We make a point of ensuring diversity among the panel by including women, people from around the globe, a variety of disciplines, and age differences. The wide range of backgrounds can be seen below to include executives of high-tech companies, government researchers, academics, consultants, forecasters, and futurists. The caliber of our authorities is often unsurpassed, but not all experts must be famous to be included. These guidelines help provide a varied group that collectively represents the best knowledge available in science and technology.

The responsibility of experts is to review the scanning analyses online, and use their best judgment to provided forecasts for roughly a dozen technologies in seven fields. Not all experts respond to each item, so sample sizes typically run in the 50 to 60 response range, which is more than ample. In return for their service, experts receive a free subscription, prominent visibility, consulting work, professional contacts, and an opportunity to participate in the best online technology forecasting system. For more details see www.TechCast-org/About/Method.

Denis Lima Balaguer, MSc
Professor of Strategic Management and Innovation Management
Vale do Paraiba University, Brazil
denis.balaguer@gmail.com

Enric Bas, PhD
Professor of Social Foresight
Department of Sociology II, Faculty of Economics
University of Alicante, Spain
bas@ua.es

Harvey Bazarian, CEO
Bazarian Technology Consulting, USA

Hellmuth Broda
European CTO
Sun Microsystems, Germany

Lynn Burton, PhD
Chair, Humanties Department
Simon Fraser University, Canada

Dennis Bushnell
Chief Scientist
NASA Langley Research Center, USA

Kemal Cakici, PhD
Lead Systems Analyst
Anteon Corporation, USA

Kelly Carnes, JD
President and CEO
TechVision21, USA

Arun Chainani, MD
Senior Advisor
Strategy & Corporate Development, USA

Andrew Chan Yik Hong
Managing Consultant
IBM Global Business Services, Malaysia

Jose Cordeiro
Founder, World Future Society, Venezuela
Chair, The Millennium Project, Venezuela
www.cordeiro.org

Donnelly Donnelly, PhD
George Washington University, USA

Jerald Feinstein, PhD
The MITRE Corporation
jerry@gwu.edu
http://home.gwu.edu/~jerry

Robert Fisher
President
The Consortium International, USA

Arnold Foudin, PhD
Chief Agriculturalist
U.S. Department of Agriculture, USA

Thomas Frey
DaVince Institute, USA

Leon Fuerth
Research Professor of International Affairs
Eliott School of International Affairs
George Washington University USA

John Geraghty
Senior Principal Information Systems Engineer
The MITRE Corporation, USA

Adam Gerber
PhD Candidate
Conservatoire National des Arts et Métiers, Paris

Jerry Glenn
Director
Millennium Project
World Federation of UN Associations

Theodore J. Gordon
Senior Research Fellow and
Co-founder of the Millennium Project
World Federation of United Nations Associations

Sam Gordy
Vice President
SAIC Corporation, USA

Janice Graham, PhD
President
Janice M. Graham Inc. USA
703.608.9421 Ext.

Martin Guyotte
US Navy War College, USA

William Halal, PhD
Professor Emeritus of Science, Technology & Innovation
George Washington University, USA
Halal@gwu.edu
http://home.gwu.edu/~halal

Ken Harris
Chairman
The Consilience Group, LLC, USA

Aharon Hauptman, PhD
Senior Researcher
Interdisciplinary Center for Technology Analysis and Forecasting
Tel-Aviv University, Israel
haupt@post.tau.ac.il
www.ictaf.tau.ac.il

Henry Heilbrunn, MBA
Shapiro Fellow
School of Media and Public Affairs, GWU, USA
Heilbrunn@aol.com
www.interactivedirections.com

William Herman, PhD
Deputy Director
Office of Science & Engineering Laboratories
Federal Drug Administration, USA

David Herrelko, PhD
Bernhard M. Schmidt Chair
University of Dayton
Retired General, USAF

Brad Hughes, MS, MBA
Director
Qwest Corporation, USA
brad.hughes@qwest.com

Sohail Inayatullah, PhD
Professor, Futures Studies, Social Sciences
Tamkang University, Taiwan

Lester Ingber, PhD
Lester Ingber Research, USA
ingber@alumni.caltech.edu
http://www.ingber.com

Michael Jackson, PhD
Chairman
Shaping Tomorrow, U.K.
mike.jackson@shapingtomorrow.com
07966 155912 Ext.

Nikhil Jagga
Systems Analyst
Accenture Corporation, USA

Niilo Kaartinen, MD
Managing Director
KRI Kaartinen Tutkimus OY, Finland

Guy Kemmerly
NASA Langley Research Center, USA

Peter King, PhD
Environmental Consultant, Thailand
Formerly Asian Development Bank

Angela Klein, MS
Assistant Professor of Computer Science
William Jewell College, USA

Jeff Krukin, MS
Executive Director
The Space Frontier Foundation, USA
jeff@jeffkrukin.com
www.jeffkrukin.com

Michael Kull, PhD
President
Amplifi Inc, USA

Young Hoon Kwak, PhD
Associate Professor
Department of Decision Sciences, School of Business
The George Washington University, USA
kwak@gwu.edu
http://home.gwu.edu/~kwak

John A. "Skip" Laitner, MA
Senior Economist for Technology Policy
American Council for an Energy-Efficient Economy, USA
jslaitner@aceee.org
www.aceee.org

Richard Lamm
Former Governor of Colorado
Director, Public Policy & Contemporary Issues
University of Denver, USA

Al Leedahl, MS
President
Engineering Design Concepts, USA
http://www.leedahl.com/engineering/design/concepts.htm

Xin-Wu Lin
Director of Research Division III
Taiwan Institute of Economic Research, Taiwan
xinwu.lin@tier.org.tw

Bob Locascio
Strategy Development Consultant, USA

David Luckey, MA, MS, MBA
Operations Analyst and Subject Matter Expert
Engineering Documentation Systems, Inc., USA
luckeyds@aol.com

Tim Mack, JD
President
World Future Society, USA

Michael Mainelli, PhD
Professor of Commerce, Gresham College
Chairman, Z/Yen Group Limited, U.K.
michael_mainelli@zyen.com
http://www.zyen.com
+44 207-562-9562 Ext.

Dr. (Econ.) Mika Mannermaa
President
Futures Studies Mannermaa Ltd., Finland

Joe Martino, PhD
Principal
JPM Associates, USA

Dennis McBride, PhD
President
Potomac Institute for Policy Studies
Research Professor, Krasnow Institute for Advanced Study, George Mason University, USA
dmcbride@PotomacInstitute.org
www.potomacinstitute.org

John Meagher, BS
International Center for Environmental Technology, USA

Danila Medvedev, PhD
General Director
KrioRus, Russia
+7 905 768-0457
danila.medvedev@mail.ru

Michael Michaelis
President
Partners in Enterprise, Inc., USA
zmichael@verizon.net
301-986-1950
301-986-1501 fax

Riel Miller, PhD
President
XperidoX Consulting, USA
rielm@yahoo.com

Steve Millett
Partner
Social Technologies Inc., USA
steve.millett@soctech.com

Patrick Moorhead
Vice President
AMD Corporation, USA

Phillip Nelson, PhD
Acceleration Studies Foundation, USA
penelson@belf.org

Eric Newman
System Architecture
AOL Corporation, USA

Bruce Niven, M. Eng.
Director
Advent Technologies
ITOCHU Aviation Inc., USA
bniven@advent-tech.com

Amy Oberg, MS
Corporate Futurist
Kimberly Clark Corporation, USA

Bob Olson
Senior Fellow
Institute for Alternative Futures, USA

Erik F. Øverland
Expert on Foresight and Innovation Policy
The Ministry of Education and Research, Norway
erik.overland@kd.dep.no or erik.overland@subito.as
www.subito.as
95963817 Ext. 0047

David Passig, PhD
Head, Graduate Program in ICT & Education
Bar Ilan University, Israel
http://www.passig.com

Ian Pearson
Futurologist
BT Group Chief Technology Office, USA
http://www.btinternet.com/~ian.pearson

Jonathan Peck
President
Institute for Alternative Futures, USA

Leonardo Pineda, PhD
CEO
Qubit Cluster, Mexico
http://www.qubitcluster.com

Joanne Pransky
Robotics Consultant, USA
http://www.robot.md

James Richardson
Futurist, USA

Carlos Ross, PhD
President
Proa Consultores, Mexico

Harry Rothman, PhD
Editor
Technology Analysis & Strategic Management, U.K.
www.tandf.co.uk/journals/titles/09537325.asp

Herb Rubenstein, JD
Chief Operating Officer
International Center for
Appropriate and Sustainable Technology, USA

Steve Ruth, PhD
Professor of Information Systems
George Mason University
International Center for Applied Studies in IT, USA

John Sagi, PhD
Associate Professor of Business
Anne Arundel Community College, USA
http://www.ola4.aacc.edu/jsagi/

Walid Saliby
Director
International Project Management
AOL Corporation, USA

Alison Sander, JD/MBA
Consultant, USA

Miriam Sapiro, JD
President
Summit Strategies International, USA

Carlos Scheel, PhD
Professor of Technological Innovation
EGADE, Tecnologico de Monterrey, Mexico
cscheel@itesm.mx

Jerry Schneider, PhD
Professor Emeritus
University of Washington, USA

Torsten Scholl, MBA
CEO, omniwatt AG, Germany
torsten.scholl@omniwatt.de

Chadwick Seagraves, MS
Library Systems Analyst
Indiana Cooperative Library Services Authority, USA
infosciphi@gmail.com
http://infosciphi.info

Yair Sharan, PhD
Director
Interdisciplinary Center for Technological Analysis & Forecasting
Tel Aviv University, Israel
sharany@post.tau.ac.il

Art Shostak, PhD
Professor of Sociology Emeritus
Drexel University, USA

Hugues Sicotte, PhD
Scientist
SAIC Corporation, USA

S. Fred Singer, PhD
President
Science & Environmental Policy Project, USA

Dexter Snyder, PhD
Technical Fellow
Technology Intelligence Directorate
General Motors Corporation, USA
dexter.snyder8212@sbcglobal.net

Knut Solem, PhD
Professor of Environment, Technology, and Social Change
Norwegian University of Science and Technology, Norway
www.svt.ntnu.no/iss/knut.erik.solem/card/index.html
+47 73 59 66 65 Ext.
+47 73 59 15 64 fax

McDonald Stewart, MS
Consultant, USA

Roger Stough, PhD
Associate Dean
School of Public Policy
George Mason University, USA

Xi Su, PhD
VP of Technical Operations
US China Business Solutions, Inc., China
xisu@uschinabiz.com
www.uschinabiz.com
202-204-3055 Ext.
202-204-3056 fax

Micah Tapman, MBA
President
Quest Consultants, USA
http://www.tapman.net

Wesley Truitt, PhD
Executive-in-Residence
College of Business Administration
Loyola Marymount University, USA

Arch Turner, PhD
Booz-Allen Hamilton, USA

Richard Varey, PhD
Professor of Marketing
Department of Marketing
The Waikato Management School, New Zealand
rvarey@mngt.waikato.ac.nz
http://www.mngt.waikato.ac.nz

Forrest Waller, MPA
Senior Scientist
SAIC Strategies, USA

Judson Walls, MBA
Associate
Booz Allen Hamilton, Inc., USA

Ann Wang, PhD
Assistant Professor
University of DC, USA

George Whitesides
Executive Director
National Space Society, USA

Name Index

Subject Index